EXTREME

EXTREME

why some people thrive at the limits

EMMA BARRETT | PAUL MARTIN

OXFORD
UNIVERSITY PRESS

OXFORD
UNIVERSITY PRESS

Great Clarendon Street, Oxford, OX2 6DP,
United Kingdom

Oxford University Press is a department of the University of Oxford.
It furthers the University's objective of excellence in research, scholarship,
and education by publishing worldwide. Oxford is a registered trade mark of
Oxford University Press in the UK and in certain other countries

© Emma Barrett and Paul Martin 2014

The moral rights of the authors have been asserted

First Edition published in 2014

Impression: 1

Published in the United States of America by Oxford University Press
198 Madison Avenue, New York, NY 10016, United States of America

British Library Cataloguing in Publication Data
Data available

Library of Congress Control Number: 2014931568

ISBN 978-0-19-966858-8

Printed in Italy by L.E.G.O. S.p.A.

Links to third party websites are provided by Oxford in good faith and
for information only. Oxford disclaims any responsibility for the materials
contained in any third party website referenced in this work.

ACKNOWLEDGEMENTS

We are very grateful to the following people for their help and comments: Patrick Bateson, Lindsey Chiswick, Jill Farrow, Andrew Glazzard, Latha Menon, Emma Ma, and an anonymous reviewer.

CONTENTS

ABBREVIATIONS

BPS	Boredom Proneness Scale
CBS	Charles Bonnet Syndrome
DORA	dual orexin receptor antagonists
fMRI	functional magnetic resonance imaging
HP	harmonious passion
HPA	hypothalamus-pituitary-adrenal
NDM	naturalistic decision making
OIC	officer in charge
OP	obsessive passion
PSS	Preference for Solitude Scale
PTSD	post-traumatic stress disorder
REM	rapid eye movement
REST	Restricted Environmental Stimulation Technique
RPDM	Recognition Primed Decision Model
SAS	Special Air Service
SSS	Sensation Seeking Scale
STAI	State-Trait Anxiety Inventory
WNSS	Weinstein Noise Sensitivity Scale

1

Life at the Edge

Out on the edge you see all kinds of things you can't see from the center.
Big, undreamed-of things—the people on the edge see them first.

Kurt Vonnegut Jr, *Player Piano* (1952)

Throughout human history, some people have chosen to put themselves into extreme environments. On land and sea, at the poles, in the air, and in space, they place their lives at risk in order to work and play in environments that expose them to the most intense physical and mental demands.

This book examines extraordinary human endeavour and the psychological qualities that underpin it. Drawing on real-life cases, including those of explorers, mountaineers, deep-sea divers, and astronauts, we explore their personal characteristics and complex motivations, and analyse the psychological attributes that lead to success (or failure) for individuals and teams.

We draw on scientific research to understand what happens, both mentally and physically, to people at the limits of human experience. What psychological and emotional qualities does a person need to survive and thrive in such hard places? How do they prepare for situations that are beyond the limits of normal experience? What are the long-term effects on them and their families? And why on earth do they choose to do it in the first place?

We also consider the practical lessons we can all learn from understanding how people cope in hard places. Being brave, making good decisions, planning and preparing, dealing with social conflict, working in small groups, learning to focus attention, coping with boredom, sleeping well, and building psychological resilience are valuable skills in everyday life, just as they are in extreme environments.

We begin by taking a look at some of the hardest of hard places, and the people who choose them.

Hard places

Apsley Cherry-Garrard described polar exploration as 'the cleanest and most isolated way of having a bad time which has been devised'.[1] Some 8,000 men applied to join the 1910–1913 British Antarctic expedition led by Captain Robert Falcon Scott. The handful who made it through the selection, including Cherry-Garrard, willingly subjected themselves to danger, snow blindness, frostbite, exhaustion, dysentery, hunger, and of course extreme cold. Many did not survive.

The winter before Scott's ill-fated walk to the South Pole, Cherry-Garrard and two other members of Scott's team embarked on a journey to collect Emperor penguin eggs.[2] During the black depths of the Antarctic winter, the three men hauled two sledges weighing a third of a ton across the frozen wilderness in complete darkness and temperatures of minus 60 degrees Celsius. Their rations consisted only of pemmican (a paste of meat and fat), biscuits, butter, and tea, and they had no means of communicating with anyone else.

During their five-week mission, Cherry-Garrard and his two companions often came close to death. In the pitch-blackness, each of them frequently fell into deep crevasses from which they had to be hauled by the others. With perhaps just a hint of understatement, Cherry-Garrard noted that: 'Crevasses in the dark *do* put your nerves on edge'.[3] The cold was so intense that their teeth split and they became frostbitten even when inside their sleeping bags. The suppurating blisters on his fingers made even the stoical Cherry-Garrard howl with pain. Against all odds, though, they reached their destination and returned with their prize of three penguin eggs.

A decade later, Cherry-Garrard observed that the Winter Journey, as it came to be known, 'had beggared our language: no words could express its horror'. His suffering had been so great that he had hoped only to die without much pain. It was, as he put it, 'the weirdest bird's-nesting expedition that has ever been or ever will be'.[4]

Climbing mountains is another extreme activity in which people expose themselves to intense hardship and risk. Errors of judgement and bad luck can end a climber's life in many different ways, including by falling, avalanche, rock fall, altitude sickness, or freezing. Europe's highest mountain, Mont Blanc, has alone claimed more than a thousand lives.[5]

In 1936, four German and Austrian mountaineers attempted to be the first to climb the fierce North Face of the Eiger. Reaching the summit requires a technically demanding piece of climbing across an apparently impassable slab of sheer, icy rock. During the ascent, Andreas Hinterstoisser was the first to cross the 'Hinterstoisser Traverse', as it was later named in his honour. His companions followed, using the rope he had set up, retrieved the rope, and continued their ascent.

All went well until falling rocks injured Willy Angerer, forcing the group to abandon their summit attempt and descend to the Hinterstoisser Traverse. Only then did they realize that it could not be crossed in reverse without fixed ropes. With the Traverse impassable, they were forced down a much more perilous route. As they descended, an avalanche struck them. Hinterstoisser fell to his death, Angerer was propelled into the mountainside with fatal force, and Edi Rainer was asphyxiated in his ropes. The remaining climber, Toni Kurz, was left dangling at the end of his rope, within sight, but not reach, of a rescue party that had struggled for hours to reach him. He died within a few arms-lengths of rescue. A cascade of bad luck had created a situation that was beyond the climbers to overcome.[6]

Equally daunting hazards, including drowning, asphyxiation, ruptured lungs, and decompression injuries (the 'bends'), face those who dive deep under water. To avoid the bends, divers must pause for long periods during their return to the surface, a process that can take many hours on the deepest dives.[7] During these decompression stops, the diver must cope with cold, dehydration, and sheer tedium. And the stops do not always go smoothly: one cave diver almost died when a gas bubble formed in his inner ear during an ascent from 250 metres, destroying his sense of balance and making him uncontrollably nauseous. He was completely disorientated, vomiting repeatedly, and barely able to hold on to his guideline. Nonetheless, he still had to complete more than 10 hours of decompression before reaching the surface.[8]

Cave diving is a particularly dangerous form of diving. It has been likened to swimming down a flooded lift shaft in a ten-storey office block in complete darkness. The cave diver can easily become lost or trapped, and merely touching the floor can stir up a cloud of silt that cuts visibility to zero, making it impossible to distinguish up from down.

When death does come in a cave diving accident, it does not always come quickly. In one case, a diver became lost while exploring a labyrinth of flooded caves in South Africa. Rescuers found his body six weeks later on a dry rock shelf. He had climbed out of the water and survived for three weeks before eventually dying alone in absolute darkness.[9]

The pursuit of extreme activities has taken people to great heights. Twelve men risked everything to stand on the surface of the Moon. Hundreds more have braved the vacuum of space. Scott Carpenter, who became the fourth American in outer space when he orbited the Earth in 1962, said: 'I volunteered for a number of reasons. One of these, quite frankly, was that I thought this was a chance for immortality. Pioneering in space was something I would willingly give my life for'.[10] Although Carpenter eventually succumbed to old age, eighteen people have died during space missions, including three cosmonauts who asphyxiated when their air supply leaked into space as they undocked from a space station.[11] Sending people into space—and returning them safely—is one of the ultimate tests of technology and teamwork.

Others who explore the skies include test pilots such as Bill Waterton, who in 1952 took off in a prototype jet aeroplane minutes after witnessing a fellow pilot die in a horrific crash. Waterton, who still had burn marks from a crash he had survived nine days earlier, was forced to taxi along one edge of the runway to avoid the smouldering wreckage of his friend's plane.[12] The ability to overcome fear and manage anxiety is a prerequisite for coping with extremes.

Despite the high mortality rate and meagre pay, the life of a test pilot in the decade following World War II offered extreme sensations that few humans ever experience. One pilot rhapsodized about taking a new type of jet fighter up to 40,000 feet and pointing it vertically downwards on full throttle: 'Soon you are going straight down at the earth at supersonic speed. You can see the earth rushing up towards you. It's a wonderful thrill'.[13] (He later fractured his spine after a crash-landing.) Extreme environments clearly offer the prospect of rewards that are unavailable in everyday life.

Some people travel to extreme altitudes in order to plummet back to Earth. The record for the highest parachute jump is currently held by Felix Baumgartner, who in 2012 stepped from a balloon 24 miles above the Earth. At such altitudes a human body would suffer catastrophic damage unless heavily protected. Without a pressure suit, bubbles would form inside the body, blocking arteries and causing extreme pain. The lungs would haemorrhage because of the pressure difference. Even with a pressure suit, a high-altitude parachutist faces tremendous hazards. A free fall can potentially reach the speed of sound, and there is the danger of succumbing to an uncontrollable 'flat spin' of up to 200 revolutions a minute.[14] Managing such extraordinary risks requires meticulous preparation and extensive practice.

Then there are those who brave the oceans. Even a circuit of the British Isles can be lethal, as competitors in the 1979 Fastnet Race discovered. Two days into the race more than 300 yachts were hit by a huge storm and fifteen sailors died. Two more sailors were left for dead on their sinking yacht, one with severe injuries. Anxiety, dehydration, pain, and exhaustion impaired their ability to think clearly about how to deal with their situation. They were rescued in the nick of time. More than 4,000 people took part in the Fastnet rescue, one of the largest in sailing history, and three rescuers died.[15] Pursuing extreme activities can put other people's lives at risk as well.

Stories like these underline the physical and psychological demands that characterize extreme environments, and the sorts of qualities people must possess if they are to survive in them. An early Antarctic explorer captured several common themes when he wrote of 'the strain of the whole thing, the exhaustion and pain, the cold, the want of food and sleep, the monotony, and the anxiety as to what is to happen at the end'.[16]

What all extreme environments have in common, according to one established definition, is making 'extraordinary physical, psychological, and interpersonal demands that require significant human adaptation for survival and performance'.[17] As this definition makes clear, the demands are not just physical. An individual's psychological response is often the most important factor. The ability to cope with stressful situations depends at least as much on the mind as it does on the body.

Stress and stressors

Extreme environments are undoubtedly stressful. But what does that mean? Stress occurs when an individual is subject to demands that exceed, or threaten to exceed, their capacity to cope.[18] As such, stress depends on both the environment and the individual. The extreme environments described in this book have features that make them objectively demanding by any standards. But the extent to which they are stressful will depend on the individual's ability to cope with those demands, which in turn depends on factors such as their skills, experience, and physical and mental state at the time.[19]

A couple of definitions may be helpful. The unpleasant and potentially harmful stimuli that are capable of causing stress are often referred to as stressors, while the psychological and physiological reactions they elicit are known as the stress response. As we shall see, stressors may be physical, psychological, or social.

Stressors vary in their impact according to factors such as their severity, duration, predictability, and controllability. Other things being equal, a stressor will have a bigger impact, both psychologically and physiologically, if it is persistent and uncontrollable—in other words, if it lasts a long time and there is little you can do to avoid it.[20] Stressors encountered in extreme environments often have both characteristics.

What does stress do to us? The answer depends on a number of things, including whether the stressor is acute (short-lived) or chronic (prolonged). The immediate response to an acute stressor is sometimes referred to as the fight-or-flight response. It involves rapid adjustments in cognition (thinking) and physiology that prepare us for a challenge which threatens our well-being. Energy reserves are mobilized and attention is sharply focused on the immediate threat. The pupils dilate to let in more light and reaction times speed up. If the threatening event does materialize, the brain stores vivid memories of the experience, enabling the individual to respond faster if confronted with a similar threat in future.[21]

Acute stress is accompanied by changes in brain function, in which complex thinking is reduced in favour of quick reactions. These changes speed up responses but they can also impair the ability to make complex decisions.[22] Stressful situations tend to promote short-term thinking. People under stress often feel pressured to take decisions quickly, even if, in reality, they do not need to. They may ignore relevant information or attend only to information that confirms their expectations. They tend to consider fewer options and fall back on tried and tested responses. Stress can produce a form of mental paralysis, referred to as decision inertia, in which the individual procrastinates, performs important tasks slowly or not at all, and makes excuses for inaction. One of the most powerful defences against poor decision-making under stress is expertise acquired through learning and practice[23] (we explore this theme in Chapter 9).

The physiological processes underlying the acute stress response have been investigated over many decades and are reasonably well understood. Acute stress triggers a complex cascade of neural and hormonal signals, releasing a cocktail of chemical messengers that affect organs throughout the body.

Central to the acute stress response is the release of the hormones adrenaline and noradrenaline into the bloodstream. This prepares the individual for immediate action by triggering rapid rises in blood pressure, pulse rate, and breathing. The heart beats faster and pumps more blood with each beat, while blood vessels supplying the muscles expand to improve the supply of oxygen. The bronchial tubes dilate to allow more air to pass with each breath. Sweat glands are activated and blood is shunted away from the extremities of the body into the muscles, heart, and brain. The constriction of peripheral blood vessels produces the 'cold feet' (and cold hands) associated with acute stress. Digestive processes slow down, with associated symptoms of dry mouth, loss of appetite, and churning guts (hence 'getting the wind up').[24]

Another core element of the stress response is the hypothalamus-pituitary-adrenal (HPA) axis, comprising the hypothalamus region of the brain, the pituitary gland (at the base of the brain) and the adrenal glands (located next to the kidneys). Under instructions from the brain, the adrenal glands release the hormone cortisol into the bloodstream. One of the main functions of cortisol is to mobilize the body's energy reserves. The HPA axis mediates the longer-term response to stress and is activated by prolonged or repeated stressors, especially those involving social threats or negative emotions.[25]

An acute stress response may be triggered by the *anticipation* of a stressor, possibly long before it actually happens. Some time before a person makes their first ever parachute jump, for example, their adrenaline and noradrenaline levels, blood pressure, and heart rate are likely to rise, and they may experience the characteristic symptoms of dry mouth, cold extremities, and loss of appetite. These responses typically disappear soon after they have landed, when the pre-jump feeling of anxiety is succeeded by one of elation.[26]

The stress response has evolved to help us survive. But too much stress can be detrimental to physical and mental health. Traumatic experiences may leave mental scars in the form of post-traumatic stress disorder (PTSD). Chronic stress can have other insidious effects. One of its harmful consequences is altering the functioning of the immune system, making the individual more susceptible to physical illness. This happens partly because of chronically elevated levels of cortisol, brought about by activation of the HPA axis. Cortisol has a variety of biological actions, one of which is suppressing elements of immune function. Prolonged activation of the

HPA axis is associated with a range of mental and physical health problems, including anxiety disorders, depression, and cardiovascular disease.[27]

Coping with stressors

In this book we will be looking at how people cope (or fail to cope) with the multiple stressors they encounter in extreme environments. Many of these stressors are physical, such as extremes of temperature or pressure, lack of oxygen, or noise. Those who enter extreme environments must also cope with hardships such as squalor, pain, thirst, and bad sleep. As we shall see, all of these stressors have psychological and emotional consequences, including affecting the ability to think and make good decisions.

In addition to physical stressors, people in hard places must cope with psychological and social stressors such as anxiety, fear, and the pressure of living in enforced proximity to a few other individuals for long periods. Social tensions may manifest themselves as conflict between members of a team, between team members and their leaders, or between the team and 'mission control'. Lack of privacy can amplify interpersonal irritations, and poor communication makes matters worse by fostering misunderstanding. Stress can be infectious: transmitting our own anxieties to other people increases the pressure on them. Spending time in hard places also disrupts relationships with family and friends. The damage may become apparent only after the adventurer has returned to normal life.

Despite all of this, plenty of people clearly do cope with the stress of extreme environments and some even thrive on it. How do they do it?

Individuals vary considerably in their coping ability, and therefore the stress they experience under apparently similar conditions. A situation that is highly stressful for one person may be tolerable or even positively stimulating for someone else. Someone who is about to make their first ever parachute jump, for example, is likely to be more stressed than someone who has successfully jumped many times before.

Much depends on the person's assessment of their situation and of what they might do about it. This assessment process, which is known as cognitive appraisal, is crucial to how individuals deal with stressful situations. The assessment may be conscious and deliberate—as, for example, when pausing to think through the practical implications of worsening weather conditions. It may also be influenced by feelings that are below conscious awareness, as in a sense of foreboding.[28]

The behavioural and psychological responses that people use for dealing with stressors are often referred to as coping strategies. These strategies may be directed either at the stressor itself (problem-focused coping) or at the individual's emotional response to the stressor (emotion-focused coping). Strategies may involve avoiding the stressor, mitigating its effects, or tolerating the resulting stress.[29] For instance, you could respond to extreme heat with problem-focused coping strategies such as using air-conditioning, wearing suitable clothing, staying in the shade, and avoiding physical activity. You could also use emotion-focused coping strategies such as self-medicating with alcohol, distracting yourself by talking or reading, and maintaining an attitude of determined stoicism.

An individual's response to a stressful situation will further depend on factors unique to them, including their skills, experience, personality, beliefs, physical health, and experience of previous stressful events. Contextual factors, such as hunger, sleep deprivation, and social environment, also play a role.[30]

One of the most effective coping strategies is to gain a sense of control over the stressor. Individuals who feel they are in control of a challenging situation tend to be less stressed by it than those who perceive themselves to be passive victims of circumstance.

The importance of control was demonstrated decades ago in experiments in which pairs of rats, mice, or people were repeatedly exposed to identical stressors such as mild electric shocks or loud noises. One individual in each pair could control the stressor by, say, pressing a button to turn it off, while the other subject's button did nothing. Both individuals experienced the same stressor at the same time, differing only in their ability to control it through their own actions. The experiments consistently showed that an uncontrollable stressor generally evokes a bigger stress response than a controllable (but otherwise identical) stressor.[31] A person whose experience and training equip them to understand a difficult situation is better able to control it and therefore more likely to cope. In extreme environments, planning, preparation, and expertise play crucial roles in coping with stress.

Another important buffer against stress is social support. Individuals who have supportive social relationships tend to be less vulnerable to stress than those who are socially isolated.[32] However, social support is not always available in extreme environments, and personal relationships can also be a powerful source of stress.

Another way in which people cope with stressful situations is by observing how others respond and then copying them. One consequence is that particular coping strategies tend to spread within groups or cultures. Cultural differences in coping strategies have been observed in extreme environments. Psychologist Larry Palinkas and colleagues studied more than 200 men and women from five nations who were based in Antarctic research stations during the polar winter. 'Wintering over' is a potentially stressful experience that involves months of confinement with a small group of people and little possibility of rescue should anything go wrong. People from different nations varied significantly in their responses. For instance, the Poles and Russians sought social support from their colleagues more often than did the Americans.[33]

Bestselling accounts of extreme environments have, unsurprisingly, tended to emphasize suffering, reinforcing a perception that stressful activities are inherently harmful. Stress is not all bad, however. Exposure to mild, acute stressors can be stimulating and enjoyable—as demonstrated by popular forms of entertainment such as bungee jumping and theme park rides.

More significantly, the experience of coping successfully with stressful situations can have long-term psychological benefits. A growing body of evidence shows that many people who spend time in extreme environments find it a positive experience that enhances their well-being. Coping with moderate stressors can have a psychological toughening effect, making individuals better able to cope with stressful situations in the future. People who succeed in extreme activities often emerge from their experience ener-gized, confident, and keen to face new challenges.[34]

The sense of mastery and achievement that comes from coping with extreme challenges can bring lasting satisfaction and happiness.[35] Even traumatic experiences can bring long-term psychological benefits—a phe-nomenon known as post-traumatic growth. At best, extreme experiences can leave a person mentally and physically more resilient, wiser, and more satisfied with their life.

Science in hard places

Throughout this book we refer to scientific evidence about the psychology of surviving and thriving in extreme environments. You might wonder how such evidence is obtained. By their nature, extreme environments are diffi-cult places in which to work, let alone carry out psychological research. Even so, there is a substantial body of empirical evidence.

Scientists who study extreme environments use a range of methods. One source of data is first-hand accounts and contemporary logs from exped-itions, which psychologists later analyse. Evidence also comes from obser-vational studies of people in actual and artificial extreme environments. Sometimes psychologists study people who are already engaged in extreme activities, such as those who overwinter on Antarctic stations. Government agencies, notably those involved in manned space exploration, fund scien-tists to study people in simulated missions. Volunteers are confined to artificial or real extreme environments such as caves, deserts, or underwater habitats. For readers who want to know more about how such research is carried out, we have included further detail in the Appendix.

As well as published scientific research, we have drawn upon anecdotal accounts of extreme experiences. Such accounts contain many of the same themes that feature in the research, such as the stress of being isolated with a small group of people and the debilitating effects of poor sleep. We found other themes in anecdotal accounts that were less prevalent in the scientific literature, including the crucial importance of being able to focus attention, a general lack of desire to take unnecessary risks, and the positive effects of extreme experiences.

Mainstream psychological research is also relevant. We refer later to research on topics as diverse as disgust, meditation, pain, boredom, teamwork, and expertise, all of which are relevant to understanding how people cope in extremes. We have been struck by how applicable the psychology of extremes is to everyday life. Understanding why people survive and thrive in extreme environments can provide useful lessons for us all, regardless of whether we ever venture into extremes ourselves.

In summary, extreme environments are enormously varied in their physical characteristics, but they share the capacity to produce similar forms of psychological stress. Stress occurs when the demands exceed the individual's actual or perceived capacity to cope. This means there are things we can do to mitigate stress, such as planning and preparing. Our ability to cope in demanding situations depends primarily on factors such as experience, knowledge, attitudes, and personality. Surviving and thriving in extreme environments—as in everyday life—is largely a mind game.

2

Bravery

To hell with it all, let us die cheerfully.

Frank Hurley, Antarctic explorer: private diary, January 1913

Extreme environments can be terrifying. People who venture into dangerous places face the prospect of unpleasant and potentially debilitating fear. To survive and thrive in extreme environments, they must control their fear and, in many cases, display remarkable bravery.

Conquering fear can be a rewarding experience. As one eighteenth-century climber commented, after observing the perilous lives of chamois hunters in the Alps: 'it is these very dangers, this alternation of hope and fear, the continual agitation kept alive by these sensations in his heart, which excite the huntsman, just as they animate the gambler, the warrior, the sailor'.[1] A contemporary BASE jumper made a similar observation: 'You can't even begin to try to make somebody who hasn't done it understand how frightening, how exciting, how peaceful, and beautiful that sensation is'.[2]

The men of Scott's Antarctic expedition of 1910–1913 had to draw on deep reserves of courage every day. One of the many hazards they faced was the unpredictable plunge into a crevasse, as the ice suddenly gave way beneath their feet and a seemingly bottomless pit opened beneath them. 'I had the misfortune to drop clean through, but was stopped with a jerk when at the end of my harness', wrote a member of Scott's team. 'It was not of course a very nice sensation, especially on Christmas Day and being my birthday as well. While spinning around in space like I was it took me a few seconds to gather my thoughts and see what kind of a place I was in'.[3] Another team member fell into crevasses eight times in the space of 25 minutes. 'Little wonder he looked a bit dazed', commented one of his companions.[4]

People who choose to engage in risky activities may be regarded as brave, heroic, or even reckless. The philosopher Charles Carroll Everett described the relationship between bravery and recklessness in these terms: 'As the coward sees danger where there is, practically speaking, none, the reckless man does not see it where it actually exists ... The really brave man does not overlook the danger. He does not let his mind dwell upon it; but if it exists he knows just what it is'.[5] In this chapter we look at the emotions of fear and anxiety, and at the qualities of bravery, heroism, and recklessness. We start with what it feels like to be in the grip of fear.

Being fearful

Fear and anxiety have similar psychological and physical symptoms, and the terms are sometimes used interchangeably. However, psychologists often draw a distinction between the two.[6] Fear may be regarded as a state of apprehension and physiological arousal that is triggered by the presence of a specific and imminent threat to well-being. The state of fear helps us to prepare for danger, by focusing attention on the threat and boosting physical readiness to respond. Fear is an adaptive response that starts rapidly when the threat appears and ends when the threat goes away.

Anxiety, on the other hand, is a more diffuse state. It may be triggered by less specific or less tangible threats, including a general apprehension about possible future events or personal concerns. Encountering a knife-wielding mugger on a dark street would normally trigger fear, whereas a person might feel anxious about the possibility of there being muggers in the vicinity. Unlike fear, anxiety can be long lasting. Some unfortunate people suffer from an anxiety disorder: a sense of persistent, overwhelming, and distressing anxiety that can generalize to many areas of their life.[7]

Fear has three dimensions. The first dimension is physiological arousal: the fight-or-flight response to an acute stressor, which includes a racing pulse, clammy hands, and rapid breathing. The second dimension is the cognitive response, which includes appraising the situation and considering its possible outcomes. The final dimension is the behavioural response, such as avoiding or escaping whatever is inducing the fear.

The three dimensions of fear can become disconnected: for instance, you might experience a physiological reaction, such as a racing pulse, without necessarily feeling fearful or running away. However, they are usually closely linked. A cognitive response, such as assessing a situation as threatening, triggers physiological arousal. Conversely, physiological arousal can trigger a cognitive response: having a racing pulse can make us feel apprehensive even if there is no apparent external threat.[8]

Because the cognitive and physiological dimensions are so closely linked, people can control fear by concentrating on one of the components. For instance, controlling your cognitive response by remaining focused in dangerous situations can help control your physiological response, just as concentrating on reducing your breathing rate can help you to calm racing and fearful thoughts.

Fear is a response to basic threats to survival. A vivid example is what happens when someone is deprived of oxygen, as can happen at high altitude or underwater. A standard experimental method for inducing fear and acute stress is to breathe air containing a high proportion of carbon dioxide (CO_2). Inhaling just a few deep breaths of air containing 35 per cent CO_2 can induce an acute stress response accompanied by unpleasant sensations of acute anxiety, breathlessness, and feeling trapped. It also triggers a physiological response involving rapid changes in blood pressure, breathing, and heart rate, together with the release of stress hormones. The response is intense but short-lived. In a minority of people the experience provokes a full-blown panic reaction.[9]

Breathing CO_2 appears to trigger a defence mechanism that evolved to make us respond rapidly to suffocation. But responding to actual or perceived suffocation can sometimes make a situation more dangerous. One study of scuba divers found that more than half had experienced at least one episode of sudden panic in their diving career. These episodes often start with thoughts about suffocation, which may trigger a physiological response that makes someone feel as if they are actually suffocating. The behavioural response can be to ascend rapidly, sometimes causing fatal decompression injuries.[10]

People preparing for extreme situations often experience a mixture of anxiety and excitement as they consider not only the rewards of what they are about to do, but also the dangers. The mountaineer Joe Simpson, for example, wrote of the 'hollow, hungry gap' that he felt in his stomach while waiting to start the ascent of a previously unclimbed mountain.[11]

Anticipatory anxiety may disrupt sleep, leading to poorer performance. Brendan Hall, skipper of a yacht that won the gruelling Clipper Round the World Race in 2009–2010, attributed his success in part to meticulous planning. But thinking about all the potential dangers gave him sleepless nights before the start of every leg of the race.[12] Bad sleep, which is common in extreme environments, can seriously impair people's ability to function (a theme to which we return in Chapter 4).

Once someone has started the extreme activity, anxiety often dissipates. Many divers have described feeling anxious until they are in the water, at which point they start to feel more relaxed. One wreck diver put it like this: 'The strange thing about diving, at least for me, is that the most anxious moments are in the preparation. Once in the water the weight of the gear and the worries disappear'.[13] Similarly, researchers who studied the emotions of climbers during a Himalayan expedition found that anxiety was rare, even during the most difficult phases.[14] But anxiety or fear may return quickly if circumstances alter. A change in conditions, such as deteriorating weather or an equipment problem, can trigger anxiety or fear in the most confident expert.

Feelings of anxiety or fear may persist after an extreme activity has finished. Many people feel satisfaction, perhaps tinged with relief, after they have left a dangerous environment. But those who have had a close call, or witnessed traumatic incidents such as the death or injury of a companion, may experience retrospective anxiety and fear as they realize how close they came to death. They may suffer flashbacks, during which they feel emotionally as though they were living through the experience again. In its more severe forms this can develop into PTSD.[15]

How people respond in difficult situations is partly a function of their personality. As we noted in Chapter 1, individuals vary in their reactivity to stressors. One aspect of this is the personality characteristic of trait anxiety, which is a measure of an individual's general level of anxiety. People who are high in trait anxiety are more inclined to feel anxious: they are the ones who always seem to be worried about something. Conversely, those who are low in trait anxiety are the ones who rarely seem to fret about anything.

Neither extreme of trait anxiety is particularly desirable. Anxiety and fear, in common with other negative emotions, are biological defence mechanisms that evolved to help our ancestors survive in a dangerous and uncertain world. If you are excessively anxious you will see danger everywhere and have a miserable time. But if you are not anxious enough you may fail to respond to real dangers, with potentially disastrous consequences.

Another relevant aspect of personality is known as self-efficacy. This is a measure of belief in your ability to achieve goals, and depends on your perception of your ability to control and cope with difficult situations.[16] The perceived controllability of a situation is a major factor in determining how fearful someone feels, as well as how much stress they experience. Researchers who interviewed BASE jumpers noted that: 'the decision to jump is made by balancing the natural state of fear with knowledge based on personal capabilities and technical expertise'.[17]

Experts tend to be high in self-efficacy. When psychiatrists assessed the men who became the first US astronauts in the 1960s, they found them to be neither fearless nor emotionless. The astronauts were, however, confident that they had the skills and knowledge to overcome realistic threats, and were 'not given to dwelling on unrealistic ones'.[18] It was only when the astronauts felt helpless in a dangerous situation that they felt fearful: 'When you can't do anything, that's the worst time', said one.[19]

Trait anxiety and self-efficacy affect people's responses to threatening situations—in particular, whether they tend to confront or avoid threats. Research has shown that it is generally better to confront a stressful situation; coping strategies that involve avoidance are often the least effect-ive ways of dealing with stress.[20]

People with high trait anxiety are more inclined to avoid stressful situ-ations rather than deal with them. For example, a study of athletes found that those with high levels of trait anxiety tended to fall back on avoidance strategies, such as denying there was a problem.[21] Another avoidance strategy that is often ineffective in the long run is using alcohol or other drugs to relieve anxiety. Individuals who are confident in their ability to deal with a threat (that is, high in self-efficacy) are more likely to confront and deal with problems.

The relationship between anxiety and coping strategies is not simple, however. Intriguing evidence suggests that some individuals with high levels of trait anxiety may turn to extreme activities as a way of managing their anxiety. It seems that they avoid the problems and anxieties of everyday life by throwing themselves into an absorbing activity, such as an extreme sport, in which they are more likely to succeed.

Evidence in support of this idea came from a study comparing mountain-eers with practitioners of judo (considered a low-risk sport). The results showed that mountaineers with high levels of general anxiety (that is, anxiety not specifically related to climbing) experienced significantly lower levels of general anxiety once they had completed their climb. Paradoxically, the high-risk activity of climbing appeared to have a calming effect on these naturally anxious individuals. No such effect was observed in the judo comparison group. The researchers suggested that the intense focus required for climbing served to divert the climbers' attention away from their chronic anxieties on to an external and objectively threatening situation that they felt able to control.[22]

A similar theme emerges from the autobiography of Vera van Schaik, a cave diver known for chasing depth records. Van Schaik was candid about how her extreme diving helped her to cope with chronic anxieties and feelings of personal inadequacy. 'Diving', she wrote, 'was a haven that allowed me to escape from, well, me. My personal life was this world of confusion, stress and fear in which I struggled on a daily basis to cope, never mind find any form of control. Diving was a place of peace and calm in which I excelled'.[23] Another adventurer who found peace in dangerous places was the climber Wanda Rutkiewicz, the first woman to successfully scale K2. She confessed that 'people frighten me. I'm forever scared that something terrible might happen to me'. Yet in the Himalayas she felt safe and in control.[24]

Psychologists have developed ways of measuring how participants in high-risk activities manage their emotions and the degree to which they feel they can control and influence their own lives—a trait known as agency (not to be confused with self-efficacy, which is the sense of competence for a specific activity). A series of studies found that mountaineers experienced a heightened sense of agency and ability to control their emotions while they were engaging in their risky activity, compared to people who took part in low-risk sports such as basketball. The results also suggested that mountaineers had a higher desire for control in the first place, but felt unable to satisfy this desire in everyday life and therefore had to look elsewhere. Findings such as these suggest that for many people who take part in extreme activities, a central part of their motivation is the ability to control their choices and emotions, including fear.[25]

Being brave

What does it mean to be brave? Consensus about the definition of bravery has proved elusive, but most writers agree that the essence of bravery, or courage, is carrying out a dangerous act despite being fearful. Stanley Rachman, a central figure in the psychological study of bravery, described it as 'persistence in the face of subjective and physical sensations of fear'.[26]

Following this definition, a brave person is one who experiences the cognitive components of fear, such as apprehension or dread, together with the physiological components, such as rapid breathing and racing pulse, but does not complete the trio by displaying the behavioural response and running away. By the same logic, someone who feels no fear cannot be said to exhibit bravery. Fearlessness, as distinct from bravery, means not experiencing the cognitive component of fear, as well as not displaying the behavioural response.

Bravery is a surprisingly common human capacity, even if it is not frequently tested in the safe lives that most of us lead. Many people are capable of acting courageously under the right circumstances, although most underestimate their own capacity for courage.[27] However, bravery does not always generalize across situations: some people, for example, are capable of displaying great physical courage but find social situations terrifying. Others have phobias that leave them terrified by particular objects or situations, yet they are still able to behave courageously in many other contexts.[28]

Experts have had a variable track record in predicting who will be brave and who will buckle under pressure. When psychologists and psychiatrists were asked to predict how Londoners would react to the German bombing campaign in World War II, they warned of widespread panic, neurosis, and trauma. In the event, the people of London (and of other bombed cities) adapted remarkably quickly and displayed relatively low levels of fear.[29]

We are all capable of developing our own capacity for bravery. The evidence suggests that both bravery and fear are, to some extent, learned from role models. The social transmission of bravery and fear can span generations, passing from parents to children. If your parents were fearful of a particular thing or situation, then you are more likely to fear it as well. But parents who are role models of courage and self-efficacy are more likely to produce children who grow up to be brave and capable of managing risks.[30]

Bravery, like fear, can be contagious in social situations. If one person in a group exhibits fear then the fear may spread rapidly, especially if the fearful individual is a leader or role model. Seeing another soldier 'crack up' can make troops feel anxious and increase their risk of having a similar breakdown. By the same token, people are more likely to be brave if they see a colleague or leader behaving bravely.[31] An example of this phenomenon came from a study of people who fought in the Spanish Civil War. Although these combatants felt fear, they believed they fought better when observing others behaving calmly and courageously. Similarly, in a study of World

War II airmen 80 per cent reported feeling less fearful in dangerous situations if they were aware of other airmen remaining calm.[32]

An effective way of bolstering one's own bravery is to act as a role model for others. Combat veterans have reported that trying to set a brave example helped to reduce their own fears, as well as others'.[33] The social transmission of bravery is most striking when the role model is physically present. But it can happen indirectly: even hearing stories about courageous acts can strengthen resolve. Robin Knox-Johnston, the first person to sail non-stop around the world, claimed the music of Gilbert and Sullivan gave him courage, as it reminded him of his tough British ancestors.[34]

Bravery can also be enhanced through training. The better trained you are to cope with a threatening situation, the better your decisions are likely to be, and the more confidence you will have in your ability. People who regularly put themselves into dangerous situations know the benefit of 'overlearning' skills to the point where they become automatic, because automatic skills are more likely to persist in the face of fear.[35] Overlearning is a central feature of training for the armed forces and extreme sports.

Being a hero

Brave people are sometimes described as heroes. But where does bravery end and heroism begin? For social psychologist Philip Zimbardo and colleagues, the defining features of heroic action are that it must be conducted voluntarily, for the benefit of others (rather than, say, for personal gain), and at a potential or actual cost to the hero. Heroic actions must also involve risks and difficulties that put them beyond the capabilities or imagination of most of us.[36] That is quite a high bar. Heroism researchers Scott Allison and George Goethals have proposed a less restrictive definition, which works well for extreme environments: 'When we dare to leave our comfort zone, to confront adversity rather than cower from it', they write, 'then we are demonstrating heroic tendencies'.[37]

Academics who have studied the nature of heroism generally agree that it is a social construct shaped by cultural attitudes. For instance, when

American college students were asked to describe their personal heroes, the idea of self-sacrifice featured repeatedly in their answers. Many of them named family members as their personal heroes, such as a mother who had made sacrifices to send her child to university. Some 'heroes' had not done anything particularly dramatic, but were regarded as heroes because, for example, they had faced a serious illness with unusual courage.[38] Other cultures and subcultures have different criteria for 'heroic' action, however. We suspect that most people who engage in extreme activities would not describe themselves as heroes. Nevertheless, within the distinct subcultures of, say, mountaineers or divers, certain individuals are regarded by their peers as heroes.

In common with many commentators, Allison and Goethals are sceptical of the notion that heroism is innate. They argue that even if a person is born with 'the right stuff', they will develop into a hero only as a result of social and other forces acting on them. To put it another way, adversity is required to elicit heroism. Some people deliberately put themselves into situations that afford them such opportunities, such as exploring extreme environments or choosing careers in the armed forces or emergency services. In one sense, they create their own 'heroic moments'.

What is considered heroic also depends on the historical context. In the nineteenth and early twentieth centuries, some nations had a particular predilection for heroes. Many of the pioneering expeditions mentioned in this book were founded on the desire of a colonial power to claim new territory. Some of these claims were exceptionally profitable, giving access to resources and trade routes, but others had little material worth. For example, finding and laying claim to the North and South Poles was, at the time, more a matter of national symbolism than commercial or political gain.[39] Some commentators have argued that the true goal of such endeavours was to provide 'heroic moments'. As such, even disastrous failures could be viewed as heroic successes.

Scott's final expedition is a case in point. He failed in his attempt to claim the South Pole for Britain, after being narrowly beaten by the Norwegian Roald Amundsen. On his return from the Pole, Scott penned a series of diary entries and letters as he lay dying in a storm-bound tent. His account of the doomed expedition was, according to one psychologist, in keeping with a 'culture of adventure that ennobled suffering, sacrifice, and even death, just as much—if not more—than it did success'.[40] In the eyes of the public, Scott became a valiant, tragic hero whose failure to achieve the Pole was irrelevant.

Some modern writers have been less generous to Scott's memory. The historian Roland Huntford controversially portrayed Scott not as a hero but as an ill-prepared leader with a badly thought-out plan for a venture that was clearly futile from the outset. The implication is that Scott was not heroic but reckless (an interpretation that has been fiercely contested).[41] The same charge has been levelled at others who have perished in extremes. So when does courageous risk-taking or heroism become recklessness?

Being reckless

Reckless behaviour is distinguished from risky behaviour because it is not socially approved and is likely to result in harm, especially when the harm is easily avoidable by taking precautions. This distinction between risk-taking and recklessness is not clear-cut, however. Culture plays a part, with societies that value individualism over social conformity tending to be more tolerant of individuals' rights to take big risks. Such tolerance is not limitless, however, especially when extreme activities result in a wider cost to society through deaths and serious injuries.[42]

Recklessness appears to be a personal attribute that applies across a range of circumstances. The evidence suggests that individuals who engage in one form of reckless behaviour are more likely to engage in others as well. Studies have found, for example, that people who drive recklessly are also more likely to abuse drugs and practise unsafe sex, reinforcing the view that recklessness is a coherent syndrome in its own right.[43]

What makes some individuals more reckless than others? Age is one factor. Adolescents and young adults are more likely than children or older people to engage in reckless behaviour, other things being equal. Various personality traits also contribute to recklessness, including sensation seeking, aggression, impulsiveness, susceptibility to boredom, and egocentrism. We look at sensation seeking and boredom in Chapter 12, particularly their roles in predisposing some people to start engaging in extreme activities. Social influences, from family, culture, and especially peers, also play a role in inclining some people to behave recklessly.[44] One study of several hundred young adults found that the best predictors of recklessness were sensation-seeking personality, peer influence, and 'present-time perspective' (the tendency to focus on the here-and-now rather than the future).[45]

Recklessness is a concept familiar to lawyers as well as psychologists. As far as English criminal law is concerned, recklessness is characterized by being 'advertent', which means a conscious decision has been made to take an unjustified risk. This distinguishes recklessness from mere negligence, where an individual has unintentionally taken an unjustifiable risk. As far as the law is concerned, recklessness is worse than negligence but better than intentional tort, where someone deliberately sets out to harm someone else.[46]

The distinction between recklessness and negligence can be hard to make in practice, because of the inherent difficulty of deciding whether a person's actions were the result of conscious choice rather than unintended misjudgement. How can anyone be sure that an individual's rash actions resulted from their failure to appreciate a risk, rather than a conscious decision to ignore it? Even the individual concerned may not truly know what lay behind their behaviour. One of the best protections against taking negligent risks is, of course, expertise, which we consider in Chapter 9.

In sum, then, fear and anxiety have three distinct elements: physiological, cognitive, and behavioural. We can learn to manage fear (and therefore be brave) by manipulating one or more of its three elements—for example, by using relaxation techniques to reduce the physiological response, by reappraising the situation so that it seems less daunting, or by equipping ourselves with a range of possible behavioural responses. We all have a greater capacity to be brave than we sometimes appreciate.

3

Hardship

The aim of the wise is not to secure pleasure, but to avoid pain.

Aristotle (384–322 BC)

Coping with hardship and discomfort is an essential ability for anyone who wants to survive and thrive in extreme environments. As a former US Navy Special Forces operator put it, 'One of the key lessons learned early on in a SEAL's career was the ability to be comfortable being uncomfortable'.[1]

In this chapter we examine how people cope with four particular forms of hardship that are frequently encountered in extreme environments: squalor, hunger, thirst, and pain. Other common forms of hardship, including sleep deprivation, monotony, social conflict, and solitude, are explored in later chapters. They all have a significant psychological impact.

The stark discomfort of one extreme environment—the desert—is graphically portrayed by the explorer Wilfred Thesiger in his classic of travel writing, *Arabian Sands*. Thesiger was one of history's great adventurers. During World War II he operated as a secret agent in Cairo, and fought behind enemy lines in the Libyan Desert as a member of the Special Air Service (SAS). After the war he spent his life travelling, on foot or using animal transport, throughout the Middle East, Asia, and Africa.

Thesiger's most famous desert journeys took place between 1945 and 1950, when he criss-crossed the largest sand desert in the world: the Empty Quarter of Arabia, which straddles present-day Saudi Arabia, Yemen, United Arab Emirates, and Oman. At the time, most of this region was unknown to any European. Thesiger's travels took him through areas occupied by violent and warring tribes, amid terrain so desolate that even some highly experienced Arab travellers refused to accompany him.

Thesiger was repeatedly exposed to physical and psychological hardships, including squalor, hunger, thirst, pain, sleep deprivation, interpersonal tension, and boredom. His decision to travel barefoot led to considerable pain. During the day his feet burned on the desert sand, and at night they cracked from the intense cold. On one journey, he and his Arab companions covered 2,000 miles in seven months, surviving on two pints of water a day and two pounds of flour a week. They were constantly hungry and thirsty, and subject to the sustained anxiety of being pursued by raiding parties of hostile local people who wanted to kill them. They often slept badly because of the intense cold at night. On top of that, desert travel entailed long periods of monotony.[2]

Cold places are just as unforgiving. Dire hardship accompanied numerous doomed attempts to reach the North Pole in the nineteenth and early twentieth centuries. Many of these expeditions followed a similar pattern.[3] First, there would be an overconfident leader with financial backing from an industrialist, newspaper proprietor, or royalty. Then there would be a long voyage in an ill-equipped ship that would eventually be crushed in the ice or limp home barely afloat. The outward voyage would be followed by at least one winter trapped in the ice (*four* winters in the case of Sir John Ross). Finally, there would be heroic sorties from ice-locked ships across land, sea, and ice to locate the Pole. These punishing excursions lasted weeks or months and often ended in tragedy. More than 700 of the thousand or so people who took part in these expeditions died, according to one estimate.[4] Frostbite, often followed by gangrene, killed some; accidents, malnutrition, and food poisoning killed others. A few were murdered by fellow explorers.

By and large, people who choose to enter extreme environments accept the hardship as a price they must pay. For some, however, the experience of enduring discomfort and pain seems to be a motivation in its own right. In his account of the first ascent of Everest, James Morris remarked that mountaineers were motivated by pride, ambition, love of natural beauty, and mysticism. But the tougher the conditions became during the Everest climb, the more he came to believe that the climbers were also enjoying the discomfort and squalor of it all.[5] The mountaineering writer Jon Krakauer reached a similar conclusion. He wrote:

> I quickly came to understand that climbing Everest was primarily about enduring pain. And in subjecting ourselves to week after week of toil, tedium and suffering, it struck me that most of us were probably seeking, above all else, something like a state of grace.[6]

We look at pain later in this chapter, but start with one of the least glamorous features of extreme activities: squalor.

Squalor

Life in extreme environments is often dirty and smelly. Take space capsules, for example. The Gemini space programme in the 1960s required the crew of two astronauts to work for several days in a capsule no bigger than the front seats of a small car, wearing their spacesuits throughout. They had to urinate and defecate into their underwear. NASA's 'urine management system' leaked, and the build-up of oily secretions on their unwashed skin led to chafing and irritation, especially in the groin.

Astronaut Jim Lovell, who spent two weeks orbiting the Earth in Gemini VII, said it was 'like spending two weeks in a latrine'.[7] By the time US Navy frogmen opened the hatch after Gemini splashed down in the ocean, the capsule's odour was rather different from the fresh ocean breezes outside. (Despite the squalor, morale on Gemini VII remained high throughout the mission.[8]) The Russian Mir space station also had its own distinctive smell, which one veteran likened to '12 years in a sock closet'.[9]

More recent spacecraft have been equipped with toilets, but microgravity makes these remarkably difficult to use. The toilet on the Space Shuttle had an aperture of only 10 centimetres, and astronauts required training in how to use it. Mishaps were common: free-floating faeces were known as 'escapees'.[10]

Faeces are not the only unlovely items to be found floating in a space station. The authors of one monograph described in nauseating detail the human waste products that may be encountered by astronauts (or, indeed, by people living in other isolated and confined environments). In addition to urine and faeces, they include hair, flakes of skin, bits of nails, 'glandular secretions' (sweat, sebum, seminal fluid, and mucus), 'microflora and micro-bial products' (farts, intestinal bacteria, and fungi), pus, blood, and vomit.[11] In addition to human waste, a space capsule might be contaminated by spillages of food and drinks. No wonder that messy eating and poor toilet habits have been a cause of tension among astronauts.[12]

Polar exploration is another pursuit in which squalor becomes unavoid-able. When Ranulph Fiennes and Mike Stroud walked across the Antarctic, they did not wash for three months.[13] The Norwegian explorer Fridtjof Nansen fared little better. During one of his expeditions in the 1890s, Nansen and another man spent many months living in a hut in the Arctic. However hard they tried, they were unable to clean their clothes, which were so greasy that even washing in boiling water made no difference. The months of accumulated dirt caused their clothes to stick to their bodies, leading to sores. They discovered that the best method for cleaning their hands was to lubricate them with oil and warm bear's blood, then scrub it off with moss. Nansen wrote that he now understood 'what a magnificent invention soap really is'.[14] Given how precious soap was to such explorers, the members of an earlier Arctic expedition must have been devastated when their starving dogs, having devoured their masters' boots, socks, and tobacco pouch, swallowed their only bar of soap.[15]

Early polar expeditions generally started with a gruelling voyage in an overcrowded ship in which washing was almost impossible. Such was life on board *Nimrod*, the ship that took Ernest Shackleton's 1907–1909 exped-ition to the Antarctic. One member of the expedition described the living quarters as a place in which he would not house ten dogs, let alone fifteen men. It was, he wrote, 'more like my idea of Hell than anything I have ever imagined'.[16] He observed how a colleague who was suffering badly from seasickness lay in his sleeping bag vomiting while the cook handed up food from the galley below. Later in the expedition, when three men returned to *Nimrod* after several months walking across the ice, their body odour was disgusting, and they were 'the colour of mahogany with hands that resembled the talons of a bird of prey'.[17]

Apart from not washing or changing their clothes for months on end, polar explorers who got caught in blizzards had to urinate and defecate inside or very close to their tiny tents. In such situations they aimed (literally) to avoid the patch of snow they were using for their drinking and cooking water. Climbers who bivouac on a mountainside during a multi-day climb face a similar problem.[18]

Being dirty does not generally cause health problems. But it can be psychologically stressful, especially when the dirt is unavoidable and seems disgusting. Disgust is a universal human emotion, although the stimuli that trigger it are to some extent culturally moderated. Certain things disgust everyone, regardless of cultural background, notably faeces, vomit, objects contaminated with excreta (such as dirty toilets or stained bedclothes), people who appear sick, and putrid foodstuffs. These universally disgusting things provoke what has been termed 'core disgust'. The best explanation for this response is that it evolved to help us avoid disease.[19]

Beyond core disgust, cultures and individuals vary to some extent in what other things they consider to be disgusting. For instance, when researchers asked people from different countries what they found disgusting, the responses varied from 'kissing in public' (India), to 'dog saliva' and 'aphids in lettuce' (Netherlands), and 'foul language' (United Kingdom).[20]

In this context, 'culture' includes professional cultures as well as national ones. Jack Stuster noted that in Antarctic missions military personnel tended to stay cleaner than civilians, who were less disciplined and more inclined to let their standards slip.[21]

Many pioneering explorers emphasized the importance of discipline in maintaining cleanliness. Adolphus Greely, the leader of the Lady Franklin Bay Expedition of 1881–1884, believed strongly in high standards of hygiene, even when stranded on the ice during the Arctic winter. Several pages of his memoir are devoted to hygiene, including ensuring that 'the table linen, changed twice a week, was kept neat and clean, and the table always presented a tidy, creditable appearance'.[22]

Greely waxed lyrical about the importance of a regular bath, which he and his men enjoyed daily during the early stages of the expedition. He was right to encourage bathing when possible, because a hot bath can be a rare source of pleasure in an otherwise austere regime. In an Antarctic military exercise in the 1980s, an enforced reduction in the frequency of showers led to a sharp drop in morale.[23]

Perceptions of how often people should bathe or shower, however, are culturally moderated. When discussing hygiene in space, one researcher argued in 1968 that 'personal hygiene as practiced in the US today is largely a cultural fetish, actively promoted by those with commercial interests', and therefore astronauts should be trained to accept lower standards of hygiene.[24] -

Not only do cultures vary in what they regard as clean, but so too do individuals.[25] And because individuals have different standards, squeezing together people who are easily disgusted with those who are indifferent to filth can be problematic. Personal hygiene can become a predictable source of social conflict in confined environments.[26] During the Cold War, the crews of Soviet, American, and British submarines played a deadly game of cat-and-mouse in the depths of the world's oceans, secretly gathering intelligence about their adversaries. The submarines stayed submerged for months at a time and personal hygiene was of limited concern to some personnel. A member of one US submarine's crew was known as 'Animal' because he would never take a shower and delighted in tormenting his crewmates with 'stink-off' contests.[27]

Hunger

Hunger is a risk in many extreme environments, where people may have to work hard under demanding physical conditions without a reliable supply of food. A combination of harsh climate, physical effort, and bulky equipment means people need extra food to keep going—more than 7,000 calories a day in the case of polar explorers hauling sledges.[28] The food must also be of reasonable quality if morale and health are not to suffer. Malnutrition has psychological as well as physical effects, including impaired thinking and judgement. At worst, starvation has driven desperate people to commit theft, murder, and even cannibalism. Good food and occasional treats, however, can make a huge difference to mental well-being.

The Swedish explorer Salomon Andrée and his crew, who traversed the Arctic in a hot air balloon, experienced food-related highs and the ultimate low. They packed a range of delicacies including wine, port, gateau, chocolate, and biscuits, which they consumed in a feast to celebrate Sweden's Jubilee Day in September 1897. A few months later the men died on an island off Spitsbergen, most likely from botulism poisoning.[29]

Those who travel to the hardest places usually have to carry their food supply with them, along with fuel to cook it. In such cases an accident can spell disaster, as many early polar explorers found to their cost. In 1912 Australian explorer Douglas Mawson and two companions were travelling on foot and dog sledge across a previously unexplored region of the Antarctic when one of them suddenly disappeared into a crevasse, along with his sledge and six dogs. The sledge was carrying most of their food supplies.[30]

Mawson and his surviving companion Xavier Mertz were left with enough food for ten days, but they were over 300 miles from the nearest camp. The weather was so bad that it took 4 hours to pitch the tent and cook a meal. Although it was obvious there would not be enough food, the two men had no choice but to press on. As their food ran low, they started to eat the remaining dogs. Mawson's diary records how he ate 'a great breakfast off Ginger's skull—thyroid and brains'[31] (Ginger was one of the huskies). Mertz eventually became so ill that he was unable to walk, and Mawson ended up carrying him on the remaining sledge.

By now both men were suffering from severe malnutrition, which caused their skin to peel all over their bodies. Mawson remarked on the inconvenience of having to keep clearing the accumulated dead skin from their underwear and socks. Mertz became delirious and died, possibly from the toxic effects of vitamin A caused by eating dogs' liver. Mawson continued alone, with a hundred miles still to go and almost no food left. His hair and beard fell out in handfuls and he bled continuously from his fingers and nose. Finally, after thirty days of supreme endurance, Douglas Mawson reached his other companions at their winter camp.

One of the most feared manifestations of malnutrition was scurvy, caused by lack of vitamin C. Many early Arctic expeditions were blighted by it. Two-thirds of the crew of Vitus Bering's 1740 Kamchatka expedition, for instance, died from scurvy or related complaints when they were marooned on an island in what is now known as the Bering Sea.[32]

In those days, no one knew the cause of this unpleasant and potentially fatal condition. Modern understanding of scurvy was considerably improved by research during World War II on conscientious objectors who volunteered to be human guinea pigs.[33] But long before that, it was known that fresh fruit could prevent scurvy, thanks in part to surgeon James Lind's pioneering clinical trial in 1747.[34] Polar explorers were nonetheless confused by the fact that the Inuit, whose diet was notably lacking in fruit and vegetables, rarely suffered from scurvy. Alternative theories therefore continued to hold sway. Adolphus Greely argued that scurvy was fostered by 'dampness, uncleanliness, mental ennui, too strict discipline, excessive exercise, or labour, and by regular and systematic use of alcoholic beverages'.[35] Even so, Greely correctly stressed the importance of fresh meat, which is another, less well known, source of vitamin C. Fresh meat kept the Inuit scurvy-free and proved to be a lifesaver in many polar expeditions.

Scurvy is no longer a common risk in extreme environments, but nutritional deficiencies remain a problem. In space and some other extreme environments people can become malnourished even when adequate food supplies are available. Participants in a six-month mission on the International Space Station lost about 5 per cent of body weight and had reduced levels of vitamins and minerals. Astronauts may eat less because they are feeling nauseous, because they do not have time to prepare food, or because weightlessness affects the sense of taste and smell, making food taste unpleasantly strange and unappetizing.[36]

Hunger can make people irritable and moody, or worse. In a 105-day study conducted in 2009, several 'astronauts' were isolated in a simulated 'space station'. Food became a significant source of unhappiness and conflict. When some crew members lost more than 10 kilos they were given additional food, but that made them the subject of suspicion and jealousy by the others, who were described as looking at them 'with hungry eyes'.[37]

Food became a source of conflict on Commander Finn Ronne's 1947 expedition before it even reached Antarctica. As their ship sailed south, Ronne's crew hid food from each other and looted each other's caches. 'When you wanted something you ate it quickly or got none, and this led to gorging' reported Jennie Darlington, one the two women on board.[38] Crewmembers started hoarding food: Darlington used her hatbox to secrete cornflakes, chocolate bars, and biscuits. Food-related problems continued after they had landed. Members of Ronne's American crew ignored his orders not to fraternize with the British team who were already stationed there. They secretly visited the British to enjoy the pleasures of 'rum, tea, and hot scones'.[39]

For people facing starvation, the theft of food can become a deadly matter, as it was in the case of Private Charles Henry during Greely's expedition. Members of the expedition were trapped for the winter in the far north of Arctic Canada, and someone was stealing from the meagre food stores. Private Henry gave himself away when he vomited up half-digested bacon. He was placed under house arrest and confined to his sleeping bag for a month, but was later 'paroled'. The unreformed Henry was soon caught brewing alcohol and getting 'thoroughly and disgustingly drunk'.[40]

Two months later, seven men had died from starvation and the remainder were surviving on scraps supplemented by sea-kelp, shrimps, lichen, and sealskins. There had been instances of theft by other members of the party, but Henry was the most prolific, disagreeable, and audacious thief.[41] Eventually, Greely gave him an ultimatum: steal again and risk execution. Henry was caught soon afterwards thieving shrimps from a stew, and he was executed by firing squad. Two weeks later, the remnants of Greely's expedition were rescued.[42]

Thirst

Another common hazard of extreme environments is thirst. Whereas humans can survive two months without food, we die after a few days without water. Up to seven days may be survivable in ideal conditions, but a person who is sweating in a hot environment might last only a day.[43]

Our inability to last long without water has been demonstrated in many tragic incidents. In 1877, for example, a detachment of US cavalry got lost in a desert region of northwest Texas.[44] They were without water for three and a half days in the searing heat, and four of them died. Their harrowing experience was captured in letters and interviews with survivors. On the first day, when some were already suffering from sunstroke, their thirst was severe; many were faint and some fell from their horses. By the second day their thirst was uncontrollable.

An army surgeon who interviewed the survivors shortly after they were rescued wrote: 'The most loathsome fluid would now have been accepted to moisten their swollen tongues and supply their inward craving'. By now they had no saliva and were unable to eat. They chewed bread, but the pieces stuck to their teeth and palates. Even sugar would not dissolve in their parched mouths. A range of other symptoms afflicted them: muscular weakness, loss of balance, and impairments in vision and hearing. The absence of saliva made it hard to talk: their voices were weak and sounded strange. Many were delirious. As one survivor recorded: 'Prayers, curses, and howls of anguish intermingled came to one's ears from every direction'. Sleep was almost impossible, and those who did sleep had vivid dreams about banquets in which they consumed delicious drinks.[45]

The men survived by drinking their own urine and the blood of their horses. As soon as a horse died, its veins were cut open and the blood consumed. The horses were so dehydrated, however, that their blood was thick and coagulated instantly. As the official report recorded: 'they could not swallow the clotted blood, but had to hold it in their mouths, moving it to and fro between the teeth until it became somewhat broken up, after which they were enabled to force it down their parched throats'. Even so, the blood tasted delicious to the dying men, who sucked on the organs and viscera of the dead horses. They quickly developed violent diarrhoea as a result.[46]

What little urine the men produced, which was darkly coloured, was drunk with relish after being sweetened with sugar. Some also drank the horses' urine. By this stage, they were having difficulty breathing because of the lack of mucus, making it feel as though they were suffocating. They took as few breaths as possible, breathing only through their nostrils. Then, after 86 hours without water, the survivors stumbled across a lake and were later rescued.

Thirst is unpleasant, even in its milder forms. In experimental studies, volunteers who have been deprived of water for 24 hours experience an uncomfortable dryness and a tacky, almost putrid taste in their mouth. Subjectively, their main motivation for drinking is to alleviate this unpleasant oral sensation.[47] Thirst makes water, or any other fluid, seem more desirable and more pleasurable to drink. A liquid that would normally be rejected as undrinkable—even horse's blood—will seem palatable to someone who is very thirsty. Within minutes of thirst being quenched, however, the palatability drops.

Wilfred Thesiger's years of trekking across deserts gave him plenty of experience of thirst. There might be only a single well within an area of many hundreds of square miles, and that would often run dry, while the temperature could exceed 46 degrees Celsius. Thesiger and his Bedu companions could carry enough water to last twenty days, provided they rationed themselves to a quart (just over a litre) a day for each person. They ate a starvation ration, but it was thirst that troubled Thesiger the most. 'I was always conscious of it', he wrote. 'Even when I was asleep I dreamt of racing streams of ice-cold water'.[48]

What is thirst, and how does it work? To remain healthy, an average adult male in a temperate climate requires a total of about 3.7 litres of water a day, and an adult female about 2.7 litres.[49] Most of our water intake comes from drinking fluids, including coffee, tea, alcoholic, and soft drinks, but a substantial proportion (at least 22 per cent) comes from food.[50]

For those of us living sedentary lives in temperate climates, acute thirst plays a minor role in determining our water intake. We drink mostly for other reasons. Some fluids, such as soft drinks, are consumed mainly for enjoyment or for energy; some are components of foods such as milk or soup; other fluids, such as tea, coffee, and alcoholic drinks, are consumed mainly for the psychoactive drugs dissolved in them.

The picture is markedly different, of course, in a hot climate, especially when heat is combined with physical activity. Someone who is working hard in a hot place can lose 2 litres of water an hour through sweating, and consequently requires a daily intake of 6 litres or more. Fluid loss can also be a problem for people working hard at high altitude. Athletes competing in strenuous events can lose up to 10 per cent of their body weight through sweating. When that happens, thirst becomes all too apparent.[51]

Even moderate dehydration can produce measurable impairments in our mental performance, especially if we are not used to it. Someone who is dehydrated may display deficits in alertness, concentration, and short-term memory. In some individuals, dehydration causes headaches or triggers migraine. Not surprisingly, drinking can improve alertness and cognitive performance—if you are dehydrated. However, there is little solid evidence that repeatedly drinking throughout the day improves the performance of people who are not dehydrated. Neither is there any convincing evidence that we should all drink eight glasses of water a day.[52]

Pain

Extreme environments almost invariably bring discomfort and sometimes pain. In polar regions and at high altitudes, for example, the sun's ultraviolet rays can damage the cornea, causing a painful condition known as snow blindness, in which it feels as though pieces of grit are continually rubbing across the eyeball. As well as experiencing excruciating pain, the sufferer may be rendered temporarily blind. The men in Scott's South Pole expedition found temporary relief from the pain of snow blindness by tying wads of old tealeaves over their eyes. As one of them noted, a snow-blind man is blind anyway, so tying a cloth over his eyes does not make matters much worse.[53]

Two other painful hazards of polar exploration are blistered lips and 'crotch-rot', in which constant chafing causes the groin and inner thighs to be rubbed red-raw. Ranulph Fiennes suffered badly from both while he was trekking across the Antarctic:

> The sweat-salt stung my crotch sores and the renewed assault from the ultraviolet burns further damaged my lips, from which blood and pus leaked into the chin cover of my face mask....The scabs always grew together overnight, and when I woke the act of tearing my lips apart (in order to speak and drink) opened up all the raw places. Breakfast consisted of porridge oats in a gravy of blood.[54]

Tropical climates offer endless scope for pain. The nineteenth-century explorer David Livingstone was plagued throughout the later years of his African journeys by ulcerated feet with oozing sores, which made walking torture, and by bleeding haemorrhoids, which made sitting agony. Towards the end of his life he suffered from malaria, pneumonia, rotted teeth that he had to extract himself, and digestive problems caused by his inability to chew.[55]

The explorer John Hanning Speke suffered a painful episode after being overrun by beetles while sleeping in the African bush. One beetle got into his ear and, in desperation, Speke jabbed it with a penknife. The result was a perforated eardrum followed by inflammation, suppuration, and severe swelling of his entire face. Speke described it as 'the most painful thing I ever remember to have endured'.[56] Other colonial explorers of Africa were plagued by internal parasites or contracted lockjaw. And even if nature spared them, they risked being attacked by local people. According to one account, angry tribesmen castrated and dismembered one French explorer while he was still alive, before beheading him with a knife that his executioner had sharpened in front of him.[57]

Pain is a complex beast that comes in many different forms, ranging from mild headaches and the acute pain of a minor injury to the relentless, excruciating agony of chronic pain experienced in, for instance, severe arthritis. One medical authority described pain as 'our most powerful emotion, an essential learning tool, a major factor in our relationship with the world, and a source of much of our social behaviour'.[58] The accepted official definition, as set out by the International Association for the Study of Pain, is: 'an unpleasant sensory and emotional experience associated with actual or potential tissue damage, or described in terms of such damage'.[59]

Pain has three fundamental components: sensory, emotional, and cognitive. The sensory component is the initial recognition of the physical sensation, such as the stab of pain you would feel if, like Speke, you were to thrust a penknife into your ear. The sensory response is followed by an aversive emotional response, which motivates us to make the pain go away. The emotion tells us that something bad has happened. The third component is cognitive: making us worry about the causes, significance, and implications of the pain. The emotional component is what makes pain 'pain', and the cognitive appraisal is what turns pain into suffering.[60]

Pain is a subjective experience that does not necessarily correlate with the objective degree of physical injury. This was highlighted by a study of soldiers who had been injured in World War II: most of them reported feeling little or no pain at the time they were wounded, even in cases of extensive injury.[61] Conversely, some unfortunate individuals suffer agonizing and persistent pain in the absence of any detectable injury or disease.[62]

Under conditions of severe acute stress, such as those associated with combat injuries, the brain releases neurotransmitters, known as endogenous opioids, which suppress pain. As their name suggests, these neurotransmitters are related in their chemical structure and actions to opium-derived drugs such as morphine and heroin. Opioids relieve pain in part by dissociating the sensory experience from the cognitive response, such that the pain loses its capacity to cause suffering. The net result is that extreme acute stress can have an analgesic effect, temporarily inhibiting the subjective experience of pain.[63] But while acute stress can reduce sensitivity to pain, prolonged (or chronic) stress generally has the opposite effect. Indeed, some scientists believe that stress-induced hypersensitivity to pain may lie behind some persistent disorders such as irritable bowel syndrome and chronic pelvic pain.[64]

The subjective nature of pain means that the best way of measuring how much pain someone is feeling is simply to ask them. There is as yet no objective pain-measuring equivalent of a thermometer or blood pressure meter. The most widely used tool for assessing pain is a scale from zero to ten, with ten denoting the worst pain imaginable. This remarkably simple technique generates results that are clinically meaningful, reliable, and reproducible.[65]

Recent research suggests that patterns of brain activity might provide a different, and in some respects more objective, way of measuring pain. Studies using functional magnetic resonance imaging (fMRI) have found that pain is associated with a distinctive pattern of activity involving several different parts of the brain—thus confirming that there is no one part of the brain that might be regarded as 'the pain centre'. Instead, a network of different brain regions, sometimes referred to as the 'pain matrix', appears to be involved.[66]

A person's subjective experience of pain depends on a number of psychological variables, including their current emotional state, past experience, knowledge, personality, and expectations. Pain tends to feel worse if its causes are unknown, or if it arrives after anticipation and you are powerless to do anything about it—as in torture and dentistry, for example. Uncertainty about the nature or severity of anticipated pain also makes it worse. Experiments have shown that giving people information about the pain they are about to experience, thereby reducing the uncertainty, makes the pain feel less unpleasant when it does arrive.[67]

Anxiety tends to make people focus their attention on pain, making it feel worse. Conversely, almost any form of distraction tends to ease pain. One of the most effective non-pharmacological ways of alleviating pain is by generating a mild but distracting pain in another part of the body. This useful phenomenon has been exploited for centuries in techniques such as rubbing, applying hot poultices, heating, scratching or blistering the skin using irritants, or causing bruises and swelling by 'cupping'. These mildly unpleasant stimuli also engage physiological mechanisms that result in the release of endogenous opioids, bringing relief from the more severe pain elsewhere.[68] Other forms of distraction, such as talking or watching an engaging film, can work too.

For similar reasons, pain can also be eased by deliberately diverting one's attention away from the pain and focusing instead on emotionally arousing or pleasant thoughts ('going to your happy place'). However, it is usually easier to be distracted than to distract oneself. Techniques that strengthen our ability to control our own attention, such as meditation, can help to cope with pain.[69] We return to attention and meditation in Chapter 10.

Individuals differ in their tolerance of pain and suffering. For instance, Douglas Mawson is said to have had an exceptional capacity to withstand pain.[70] Tolerance of pain is also influenced by social and cultural factors. Some cultures have strong norms of bearing pain stoically, whereas others tend to be more expressive. Soldiers are trained and expected to be stoical in the face of discomfort and pain.

An individual who has suffered great pain as a result of engaging in an extreme activity might be deterred from repeating the experience. However, research has shown that our recollection and perception of past painful experiences is largely determined by two factors: how bad the pain was at its peak, and how bad the pain was when it ended. This phenomenon is referred to as the peak-end effect. It means that we tend to prefer painful experiences that end well, even if they last longer and are objectively worse than mildly painful experiences that end badly.[71]

In experiments to investigate the peak-end effect, volunteers subjected themselves to pain by holding their hand in freezing water. They did this under two different conditions: freezing water for 60 seconds; or freezing water for 60 seconds followed by a further 30 seconds of immersion in water that was one degree warmer. The peak intensity of pain was the same in both conditions, but the ending was marginally better in the second. The second condition was objectively worse, because the pain lasted longer.

Nonetheless, most participants preferred the second condition and rated it as less painful overall.[72]

As awful as it may be, pain exists for a reason. It evolved to protect us from harm. One leading pain researcher described it as an 'internal action plan' for avoiding and dealing with harm.[73] Acute pain teaches us to avoid or escape from potential causes of injury. If we are injured, pain immobilizes the damaged area to allow healing to take place. People who are unable to feel pain—a rare condition known as congenital analgesia—suffer physical harm and die prematurely from the effects of accumulated injuries.[74] As one expert put it: 'A life without protective pain is not a happy life'.[75]

The study of hardship in extreme environments highlights some themes of broader relevance to everyday life. Unpleasant sensations such as hunger, thirst, and pain exist for a reason: they have a protective function. Feeling hungry, thirsty, or dirty may be a relatively unusual experience for many people in wealthy nations, but there is nothing fundamentally pathological about it. Society tends to overestimate the harm associated with the sensations of moderate hunger, thirst, or squalor. Our ability to tolerate pain can be strengthened by knowledge, distraction, and the ability to control attention. Coping with hardship is as much about psychological as physical responses.

4

Bad Sleep

You lack the season of all natures, sleep.
William Shakespeare, *Macbeth* (1606)

Bad sleep is an endemic hardship in extreme environments. Stress and the disruptiveness of life in extremes can lead to a reduction in the quantity and quality of sleep. People in hard places may also have to sleep at odd times, in conflict with their circadian rhythms.

Sleep loss may be acute, as in a single episode of staying awake for many hours, or chronic, as in repeatedly getting only a few hours of poor quality sleep every day. Either way, bad sleep has corrosive effects on mental and physical well-being. It undermines the ability to make good decisions and perform critical tasks, which makes it dangerous as well as unpleasant. In this chapter we consider the nature and impact of bad sleep, and ways of blunting its effects.

Sleepless extremes

Most people who venture into hard places find that sleep is one of the first things to suffer. Bad sleep is common on space missions, for example. Astronauts orbiting the Earth are exposed to sixteen sunrises every 24 hours, which can disrupt their circadian rhythms. Moreover, the time allocated for sleep is often curtailed by the need to complete urgent tasks.[1]

During the early stages of Captain Scott's final Antarctic expedition, sleep was squeezed out by the heavy demands of work. For weeks on end, Scott's men did not get to bed before midnight, but were rarely allowed to sleep past

5.00 a.m. Once they were out of bed they worked hard throughout the long day. Their fatigue was such that if they sat down, even for a few moments, they immediately fell asleep.[2]

During their infamous Winter Journey to collect penguin eggs, Apsley Cherry-Garrard and his two companions battled for weeks with crushing weariness. They were so tired that they were barely conscious of ever sleeping, although they knew they must have slept because they sometimes heard one another snore or remembered a dream. Before long, they were falling asleep while marching and waking only when they bumped into each other. Cherry-Garrard observed that 'there was nothing on earth that a man under such circumstances would not give for a good warm sleep'.[3]

In extreme environments, sleep is easily disturbed by factors such as long or unusual working hours, heat, cold, humidity, noise, altitude, illness, or anxiety. Exposure to heat, for example, results in frequent interruptions in sleep, along with significant reductions in slow-wave (deep) sleep, rapid-eye-movement (REM) sleep, and total duration of sleep. Humidity makes it worse. Exposure to cold also produces frequent interruptions in sleep, a reduction in REM sleep, and a rise in stress hormone levels. Wilfred Thesiger observed how the intense cold of the desert at night made it impossible to sleep except in short snatches.[4] Atmospheric changes can also interfere with sleep. The lack of oxygen at high altitudes leads to more frequent awakening and a reduction in deep sleep, although healthy people do generally acclimatize to altitude and the quality of their sleep improves over the course of a few days.[5]

Sleep is further disrupted in extreme environments by physical discomfort and the close proximity of other people. The members of Shackleton's 1908 Antarctic expedition slept in specially designed three-man sleeping bags, which meant that each time one man moved the other two were disturbed.[6] Mountaineers may take several days to climb the world's highest peaks, during which time they must sleep while perched on tiny ledges, or even suspended by ropes attached to a sheer face. Any sleep they do get is liable to be interrupted by storms or rock falls.

Lack of sleep has always been a feature of life at sea. Admiral Lord Nelson seldom enjoyed more than 2 hours of unbroken sleep and sometimes stayed on deck all night. Like many sailors, Nelson was adept at taking short naps during the day. We consider the power of napping later in this chapter.

The travel writer Eric Newby recalled his deep fatigue during a long sea voyage in 1938: 'I had never been so tired in my whole life', he wrote, 'far too exhausted to appreciate the beautiful pyramids of sail towering above me. In the watch below I would fall into a dreamless sleep so profound that for the first hour of our next watch on deck I would be like a somnambulist but without the happy facility of avoiding obstacles'.[7] When the ship was hit by a massive storm, the sailors who were not on watch were so tired that they slept through it, even though their compartment was inches deep in water. 'As old soldiers do before an action', wrote Newby, 'they were absorbing sleep greedily like medicine and lay snoring happily in the midst of tumult'.[8]

Long-distance single-handed sailing is an extreme activity in which normal sleep is virtually impossible. With no one to share the burden, especially in the days when technology was less sophisticated, the solo sailor must wake regularly to keep things on track.

When psychiatrist Glin Bennet studied participants in the 1972 Observer Transatlantic Race, he found that they all suffered from extreme tiredness, which they variously likened to be being drunk, high, or simultaneously asleep and awake.[9] One solo sailor managed to wake himself every half-hour, day and night, for thirty-eight days. Like all tired people, the sailors made mistakes. Some of their errors were minor, such as absent-mindedly putting equipment in the wrong places, but others were serious and, in one case, led to a collision. The tired sailors had vivid dreams and experienced auditory, visual, and even olfactory hallucinations.[10]

Sleep deprivation

Chronic shortage of sleep is a significant, if rather neglected, issue in everyday life, let alone extreme environments. Large-scale studies in several countries have found that between 5 and 20 per cent of the adult population are excessively sleepy during the day as a consequence of poor sleep at night. Significant proportions of people get fewer than 7 hours of sleep a night, even under normal conditions.[11] Being in an extreme environment almost invariably makes the problem worse. For instance, studies of Shuttle astronauts found that they slept for an average of around 6 hours a day.[12]

What does sleep deprivation do to us? The most obvious symptom is a subjective feeling of sleepiness. Objective measurements confirm that there is indeed a firm correlation between sleep deprivation and sleepiness: the longer you go without sleep, the sleepier you will feel and the faster you will fall asleep when given the opportunity. Really tired people can fall asleep almost anywhere.[13]

Scientists who conduct sleep deprivation experiments often have to work hard to keep their sleepy participants awake. One scientist who worked on sleep deprivation experiments in the 1960s recalled walking the streets with exhausted volunteers to keep them awake. Even so, their eyelids would often close as they lapsed into a twilight state somewhere between waking and sleep.[14]

The aviator Charles Lindbergh, who made the first solo flight across the Atlantic in 1927, wrote a graphic description of how it feels to battle with intense sleepiness. His 33-hour flight came close to disaster on several occasions because of tiredness, rather than any mechanical problems or bad weather. As an experienced professional pilot, Lindbergh had learned that sleep is crucial, but despite this he snatched barely 3 hours of sleep before take-off. After only a few hours in the air, he felt sleepiness engulfing him. 'My eyes feel dry and hard as stones', he wrote. 'The lids pull down with pounds of weight against their muscles. Keeping them open is like holding arms outstretched without support'.[15]

The exhausted Lindbergh tried different ways to keep himself awake. Stamping his feet and shaking his body did not work. He slapped his face, but hardly felt it; he inhaled a capsule of ammonia, but smelt nothing. Twenty hours into the flight, sleepiness finally got the upper hand and Lindbergh suddenly awoke to find his plane diving towards the water. This shocked him into wakefulness for a while, but time passed and he lapsed back into a dreamlike state. With more than a thousand miles still to fly, Lindbergh began to doubt that he could possibly stay awake long enough to avoid plunging into the ocean. But the danger of the situation finally overcame his fatigue and, as history relates, he made it to Paris and became an international hero.[16]

One of the earliest scientific experiments on the effects of sleep deprivation was conducted in 1896, when scientists at the Iowa University Psychological Laboratory kept three volunteers awake for 90 hours. As the volunteers grew more sleep deprived their reactions became slower, their memory deteriorated, and they lost the ability to focus their attention. One of them was unable to memorize in 20 minutes material that he could normally have memorized in two. Another experienced visual hallucinations, in which the air seemed full of dancing, coloured particles like flying insects. Their physical symptoms included a gradual decline in muscle strength and body temperature. The adverse physical and psychological effects were reversed after a good night's sleep and the participants suffered no lasting damage.[17]

More than a century of experimental and observational research has subsequently identified a huge array of cognitive, emotional, and physiological consequences of sleep deprivation. The principal cognitive effects include impairments in attention, vigilance, reaction times, judgement, memory, communication skills, creativity, and flexibility. The cognitive performance of tired people is also more variable and inconsistent.[18] The emotional effects include deteriorations in mood, resilience, and interpersonal skills.[19] The long-term physiological effects include increased risks for developing obesity, diabetes, cardiovascular diseases, and several forms of cancer.[20]

Individuals differ in their requirements for sleep and in their capacity to cope with sleep deprivation. Moreover, people who are chronically sleep deprived grow used to the constant feeling of sleepiness and come to regard it as normal—even if, objectively, their mood and cognitive performance are affected. The evidence suggests, however, that deficits in cognitive performance are likely to start becoming significant for most healthy adults if they repeatedly get less than about 7 hours of reasonable quality sleep a night over an extended period.[21]

Sleep deprivation is especially problematic in hazardous situations, where the cognitive demands are often intense and the consequences of making mistakes may be serious. People in extreme environments often have to make decisions that require them to assimilate and process incomplete information, while keeping track of actions, ignoring distractions, and communicating effectively with others. They must be able to spot their own errors and to improvise a new plan if the current plan is not working. Sleep deprivation erodes all of these capabilities.[22] For example, an analysis of crew behaviour on the Mir space missions found that the more their sleep had been disturbed, the greater the prevalence of errors.[23]

It takes about 16 hours of continuous wakefulness for a person to start exhibiting measurable cognitive deficits, such as slower reactions, increased errors, memory problems, and difficulty maintaining attention. A study of solo sailors found significant impairments in their reaction times after races lasting from 24 to 50 hours during which they slept only briefly.[24] Tired people also score worse on many measures of creativity, including originality, flexibility, and generating novel ideas.[25] They are particularly bad at thinking flexibly when required to revise their plans in the light of new information. These are, of course, exactly the capabilities required in extreme environments.

As you would expect, the impairments in cognitive function caused by lack of sleep make us much more likely to have accidents. Sleepiness is estimated to be a contributory factor in up to a fifth of all vehicle accidents, for example, making it a bigger cause of preventable transport accidents than alcohol and drugs.[26] The evidence from experimental studies is compelling. In one study, for example, researchers used a realistic simulator to compare people's driving ability after a normal night's sleep and after a night without sleep. The sleep-deprived drivers were much more likely to make mistakes: those who rated themselves as needing some effort to stay awake were twenty-eight times more likely to have a simulated accident, and the sleepiest participants had an accident risk that was 185 times greater than those who did not feel tired. Furthermore, the sleepiest drivers were more likely to make catastrophic errors, as opposed to minor ones.[27]

Sleep deprivation makes us accident-prone for several reasons. In addition to affecting cognitive ability, it impairs motor function and lowers mood. Worse still, a person who is very tired is liable to lapse uncontrollably into short episodes of sleep known as microsleeps, which can last up to thirty seconds. Microsleeps can occur even when someone is engaged in a demanding task such as driving, and the individual is often unaware of what is happening.[28]

The effects of sleep deprivation are similar in many respects to those of alcohol, which is why very tired people tend to behave as though they were drunk. Both tiredness and alcohol hamper our ability to perform tasks that require quick reactions, attention, coordination, judgement, and self-awareness.

Experimental evidence shows that staying awake for 21 hours produces impairments in driving ability that are similar to those caused by a blood alcohol level at the legal limit for driving in the UK.[29] To make matter worse, sleep deprivation undermines self-awareness and the ability to recognize one's own impairment. Tiredness, in common with drunkenness, can give people a misplaced confidence in their own abilities. Tired people are more inclined to take risks and behave recklessly, while at the same time perceiving themselves to be less risk-prone. In extreme environments, such behaviour could prove fatal.

Sleep deprivation also affects mood and emotional resilience. Tired people are more prone to sadness or irritation, less sociable, more easily irritated, and less emotionally robust. Severe sleep deprivation can induce feelings of persecution and mild paranoia. Volunteers in some sleep deprivation experiments have become convinced that the researchers and fellow volunteers were plotting against them.[30] Such effects could obviously have serious implications for small teams operating in extreme environments.

Lack of sleep is generally bad for the cohesion and performance of teams, especially those operating in demanding circumstances. One reason is that sleep deprivation impairs communication skills: tired people are less fluent, less expressive, and less willing to volunteer information that others might need to know.[31] Tiredness may also encourage some team members to do less than their fair share of the work and rely on others to pick up the slack—a phenomenon known as social loafing.[32] Such behaviour may cause other team members to feel resentful.

The impact of sleep deprivation on a team can be hard to predict, however. Individuals respond differently to sleep deprivation, some team members may be more sleep deprived than others, and a team's overall performance will be affected more by some individuals than others. In one study, the performance of four-person military teams was measured as they monitored simulated radar screens. After 30 hours of continuous work, the teams' speed and accuracy of decision making fell significantly. However, sleepy individuals performed better when working as part of a team than they did when carrying out the same task alone.[33]

It seems that working in a team can sometimes strengthen individual motivation. This is more likely to happen in situations where individuals are working together in close proximity and where each individual makes a unique and identifiable contribution. If, on the other hand, individuals are able to slack without being noticed, then social loafing is likely to occur.[34]

A sleep-deprived leader can have a disproportionately large effect on the team's performance. This is especially true in a hierarchical team, which will tend to behave as though it is as sleep deprived as the team's most powerful member. The worst situation is one in which a hierarchical team of specialists, who are unable to substitute for one another, are required to perform creative problem-solving tasks under the command of a sleep-deprived leader. This is, of course, what often happens in teams responding to a crisis.[35]

Sleeping at the wrong times

Extreme situations often force people to sleep at unusual times, when their body clock is out of kilter with their environment. This can be highly disruptive to sleep, even if the total time available for sleeping is adequate.

Working long shifts at odd times is a feature of life for astronauts and ground crews. In one simulation of a 105-day space mission, crew members and mission controllers worked several 24-hour shifts, during which both groups showed significant reductions in alertness, sociability, happiness, and energy.[36] More generally, people who routinely work at night, or on varying shifts, are at greater risk of suffering from chronic sleep deprivation, impaired performance, and related health problems.[37]

Shift work disrupts the circadian rhythms that evolved over millions of years to adapt humans and other animals to a 24-hour cycle of light and darkness. Normally, our levels of alertness and performance are at their peak in the late afternoon and fall to their lowest during the small hours of the morning. This means that people working at night have to sleep during their peak and work during their low point.

Extensive evidence shows that shift workers sleep less well than people who work normal daytime hours, and the impact is greatest for those who work night shifts. Shift workers spend less time asleep, on average, and feel less alert when they are awake. One study, for example, found that shift workers slept for an average of 36 minutes less each day, compared to daytime workers.[38]

Shift work is sometimes likened to jetlag, but whereas international travellers recover once they have adjusted to their new time zone, shift workers on varying schedules are persistently out of phase with their environment. Our circadian rhythms can adjust to changes in time zone at a rate of about 1 hour per day, if we are exposed to daylight at the right times.[39] However, night workers are exposed to light in the early morning, at the end of their shift, making it harder for their body clock to adjust. For that reason, people often adjust better to shift work when there is little natural light, such as overwintering in polar stations or on North Sea oil platforms.[40]

Even when shift workers sleep for the same amount of time as others, they suffer cognitive effects similar to those caused by sleep deprivation. Compared with people working conventional hours, night shift workers have reduced alertness, impaired performance, and more accidents. Many catastrophic industrial accidents have occurred at night or in the early hours, when the performance of those responsible has been at low ebb. Examples include the disasters at the Chernobyl nuclear power station in 1986 and the Bhopal chemical plant in 1984.[41] Sleep-related vehicle accidents also peak in the early hours of the morning.

Sleeping at the wrong times can also be bad for health. A substantial body of evidence suggests that shift working, especially working at night, has adverse long-term effects, which include heightened risks of cancer, cardio-vascular disease, obesity, gastrointestinal problems, and depression.[42] The association between shift work and cancer risk is now so firmly established that the International Agency for Research on Cancer classifies shift work as a probable human carcinogen.[43]

The most common health problems associated with working unusual hours are gut disorders. Shift workers, especially those working nights, notoriously experience a much higher frequency of gastrointestinal problems such as indigestion, flatulence, bloating, constipation, abdominal pain, and heartburn. One reason may be disrupted eating patterns. Research suggests that night workers tend to consume more sugary snacks and drinks, more coffee, and less dietary fibre than day workers. Their eating patterns are also more irregular, with more snacking and greater consumption of easy-to-prepare food.[44]

Napping and caffeine

The two most common remedies for sleep deprivation are napping and caffeine, which work well in extreme environments and in everyday life. Napping is a simple and effective technique that can help to mitigate the damaging effects of inadequate sleep. Moreover, napping is safe, costs nothing, requires no special equipment, and can be pleasurable.

A nap is an intentional episode of sleep during the day, lasting anywhere from a few minutes to a few hours. (Anything shorter than a few minutes is usually regarded as a microsleep, while anything longer than about 4 hours qualifies as ordinary sleep.) A nap may be taken for different reasons: to make up for lost sleep, to prepare for a period of expected sleep deprivation, or simply for pleasure.[45]

Tactical napping—the ability to sleep efficiently in short bursts—is a crucial skill for many individuals who operate solo in extreme environments. The first astronaut to sleep in space was Gordon Cooper, during a solo Mercury flight in 1963 that lasted more than 34 hours. Cooper slept in short naps, which amounted to a total of 4.5 hours.[46] In February 2001, yachtswoman Ellen MacArthur crossed the finishing line of the Vendée Globe race after ninety-four days alone at sea. She had travelled 24,000 miles across three oceans to become the fastest woman to sail single-handedly around the globe. For three months she managed her 18-metre yacht in some of the world's roughest seas, relying for her survival on tactical napping. In total, she took 891 naps, each lasting on average 36 minutes.[47]

Many famous people who were reputed to need little sleep, including Margaret Thatcher, Thomas Edison, and Napoleon Bonaparte, were in fact practitioners of napping. Winston Churchill is commonly perceived as someone who worked extremely long hours, but in reality he was an enthusiastic napper.[48]

Extensive evidence confirms that napping significantly improves the alertness, reaction times, performance, and mood of sleep-deprived people.[49] Napping helps in long-haul aeroplane flights, where it boosts the performance of flight crews during the critical phases of descent and landing. NASA research convinced airlines to introduce planned rest periods for long-haul flight crews, reversing the traditional practice of forbidding crew to sleep during a flight.[50] Napping also helps reduce the cognitive effects of shift working: a 1-hour nap taken during a night shift improves alertness and performance, although some deficits still remain.[51]

Longer naps are generally more effective than shorter naps at boosting objective performance. However, a long nap can induce a temporary but mildly unpleasant state of grogginess on waking, known as sleep inertia. Short naps avoid the problem of sleep inertia and are surprisingly effective, especially in people who are sleep deprived. Even a nap lasting just 10 minutes can boost alertness, mood, and performance.[52] With a little practice, almost anyone can learn to fall asleep and wake up 10–20 minutes later.

Napping is good for our memory and physical health. Research has shown that daytime napping improves performance on memory tasks. This improvement is correlated with the amount of slow-wave sleep obtained during the nap. The processing and formation of memory is especially dependent on slow-wave sleep, of the sort that occurs during relatively short naps.[53] Napping also mitigates some of the adverse effects of sleep deprivation on the immune system.[54]

Although napping is probably the simplest and best way of combatting sleep deprivation, the most frequently used remedy is a stimulant drug—namely, caffeine. Other stimulants, such as Modafinil and dextroamphetamine, are in widespread use, but on nothing like the vast scale of caffeine, which has long been the world's most widely consumed psychoactive drug.[55]

Caffeine is moderately effective in countering the subjective sleepiness and decline in cognitive function that result from sleep deprivation. It stimulates brain activity, leading to rapid improvements in alertness, mental performance, and mood. Even modest doses of caffeine, equivalent to one or two cups of coffee, produce measurable enhancements in attention, problem-solving ability, memory, clear-headedness, and energy.[56]

The bad news is that caffeine interferes with sleep. People with caffeine in their bloodstream take longer to fall asleep, wake more frequently during the night, and get less sleep in total. Caffeine is particularly disruptive of slow-wave sleep, which is subjectively the most refreshing element of a night's sleep. If you ingest caffeine repeatedly throughout the day, substantial amounts will still be present in your body when you go to bed, and it will interfere with your sleep.[57]

Is caffeine better than napping? In a straight comparison, the evidence suggests that napping is more effective than caffeine at combatting some of the cognitive effects of sleep deprivation, including on memory.[58] A good option can be to combine the two: a slug of caffeine followed immediately by a short nap is a proven way of countering fatigue (the idea being that you nap before the stimulant effects of the caffeine kick in).

In extreme environments, opportunities for sleep may be few and far between, and they must be seized when they arise. But that may be easier said than done, because noise, extreme temperatures, or other disturbances may make it hard to sleep. In such cases, sleep-inducing drugs, known as hypnotics (or, in common parlance, sleeping pills), may be helpful. Hypnotics are commonly taken by astronauts and are the most frequently used medications in space.[59]

Standard hypnotics can have side effects, including impairment of memory and attention, which are not ideal under extreme conditions. During their 1982 Transglobe expedition, in which they became the first people to circumnavigate the Earth's surface through both poles, Ranulph Fiennes and Charlie Burton were stuck at one point on an ice floe in the Arctic Ocean. The noise, vibration, and anxiety created by the creaking, fracturing ice on which the two men had been floating for weeks made sleep almost impossible. Fiennes noted that: 'To sleep in such conditions might not be difficult for people with no imagination at all, but for less fortunate souls a sleeping pill would be the only way to catch up on much-needed rest'.[60] They did not take sleeping pills, however, because of the lingering effects on daytime alertness. The two men needed to stay fully alert, not least to look out for marauding polar bears. A new class of hypnotic drugs known DORAs (dual orexin receptor antagonists) appear to be effective at inducing sleep without causing cognitive impairments.[61]

Alcohol has of course been used throughout history to help people fall asleep. Solo circumnavigator Robin Knox-Johnston wrote: 'I tried to rest this afternoon but couldn't because we kept yawing. We still are. I'm going to get half-drunk tonight and make sure of some real rest'.[62] Alcohol is a mild sedative and can help to speed the onset of sleep. But it also has disruptive effects on sleep, especially REM sleep, and tends to reduce both the quantity and quality of the sleep that ensues.[63]

To summarize, sleeping badly means getting an insufficient quantity of sleep, sleep of poor quality, sleeping at the wrong times, or some combination of all three. Bad sleep is common in extreme environments and in everyday life. It has a range of adverse effects on mental and physical well-being, including impairing alertness, judgement, memory, decision making, and mood. People who have slept badly behave somewhat like people under the influence of alcohol and they are more likely to have accidents. Napping is a valuable skill that mitigates the effects of bad sleep.

Society tends to underestimate the harm caused by bad sleep and many of us subject ourselves to avoidable harm by routinely going to bed late, getting up early, and having low-quality sleep in between. Just as those in extreme environments cope better when they get good sleep, most of us would feel better if we slept better.

5

Monotony

Terror and boredom are very old friends.
Martin Amis, *The Second Plane* (2009)

Feeling tired, as we saw in the previous chapter, is a notorious hazard for those who spend time in hard places. But what happens when there is nothing to keep you awake?

Many extreme environments are characterized by the visual monotony of beautiful but unchanging vistas such as ice, water, or space. When the lights are off in a cave, or at the bottom of the ocean, there is nothing to see but absolute darkness. Other senses may be deprived too. Soundscapes may be limited to the howling of wind, the crash of the ocean, or the hum of machinery. The taste of food tends to become repetitive when supplies are limited, and smells may be unpleasant but unchanging in confined spaces. Even touch may be restricted: Fridtjof Nansen's inability to change his filthy clothes for months during one Arctic expedition led him to fantasize about the sensation of soft new clothes against his skin.[1]

A lack of varied sensory experience is not the only form of monotony faced by people in extreme environments. They may also have little to do for long periods. Their working pattern may involve bursts of intense activity followed by long stretches of tedium. Overwintering in the Antarctic involves spending months cooped up with a handful of people and nothing but darkness outside. In such conditions, the lack of diversions can lead to depression and petty squabbles.

Those who organize expeditions to extreme places have always recognized monotony as a serious problem and have looked for ways to relieve it.

We discuss these later. But first, we consider the psychological effects of the most extreme form of monotony: sensory deprivation.

Sensory deprivation

What happens if you spend six months alone in a dark cave? On Valentine's Day in 1972, geologist-turned-researcher Michael Siffre was about to find out. He kissed his wife goodbye and descended a hundred feet to the aptly named Midnight Cave in Texas. He did not emerge for six months. Those months of isolation and sensory deprivation had a profound effect on his mental and physical health.[2]

Ten years earlier, Siffre had been the sole participant in a pioneering study in which he spent nine weeks alone in a cave with no way of telling whether it was day or night. He later wrote that he had become 'a half-crazed disjointed marionette' and suffered severe physical and emotional distress.[3] If that experience had been so unpleasant, then why did he choose to repeat it, and for much longer? Part of the answer is that Siffre's missions had a scientific purpose. He wanted to discover what would happen to his biorhythms when all external clues to time were removed. In this he was successful. His solitary confinement produced new findings in the field of chronobiology and helped to advance the understanding of long-term isolation. But this scientific knowledge was won at considerable personal cost.

Compared to the freezing cave that Siffre had occupied ten years previously, Midnight Cave was a balmy 21 degrees Celcius. He lived in a nylon tent atop a wooden stage of 17 square metres. The living quarters were dominated by scientific equipment, but he had some creature comforts including a freezer, furniture, books, and a music player. NASA supplied his food and monitored the experiment as part of its research for long-duration space missions. Siffre was consuming the same diet as the Apollo astronauts.

There were no clocks in the cave and Siffre had no other way of telling the time. His assistants camped at the mouth of the cave and acted as 'ground control', turning on the lights in the cave when he woke and turning them off when he felt drowsy. When the lights were off he was in absolute darkness.

Even towards the end of the experiment, when Siffre was depressed and despairing, he continued diligently to collect data. Every day he completed a battery of physical and cognitive assessments: memory tests, mental acuity tests (similar to IQ tests), tests of manual dexterity, and so on. Throughout his stay a rectal probe recorded his body temperature and electrodes on his chest recorded his cardiac rhythm. Every night he attached additional electrodes to his body so that his sleep patterns could be recorded. The data reached the surface by means of a cable. As well as acting as a tether that effectively confined Siffre to his platform, the cable occasionally delivered painful electric shocks when lightning hit the surface above him.

Siffre had prepared for the loneliness and tedium by taking books and records. But the damp caused his record player to malfunction, and within a few weeks all his books and papers were covered with mildew.

The lack of sensory and social stimulation led to a gradual deterioration in Siffre's psychological state. At one point he even contemplated suicide, only to reject the idea because he did not want to saddle his parents with the debts he had run up financing his experiment. By the seventy-seventh day his memory had deteriorated to such an extent that he forgot things unless he wrote them down immediately. He wrote in his journal: 'What am I doing here in this silly experiment while my professional life ebbs away?'[4] He called ground control: he'd had enough. Ground control told him everything was fine. He stayed.

When he eventually emerged, Siffre was a changed man. Though many of the negative effects were temporary, some were long lasting. Three years after leaving Midnight Cave, Siffre still had memory lapses, his eyesight remained poor, and he had what he described as inexplicable 'psychological wounds'.[5] He divorced and, according to one report, retreated to a South American jungle to recover from the ordeal.[6]

At face value, Siffre's psychological deterioration was not surprising. Deleterious effects of isolation have been widely reported from observational studies in prisons and anecdotal accounts of voluntary and involuntary solitary confinement. Fictional representations of individuals in conditions of solitary confinement or sensory deprivation often end in their madness. However, the evidence is by no means clear that sensory deprivation is necessarily always a bad thing. So where does the popular conception come from?

The story starts with some extraordinary laboratory studies conducted between the 1950s and the 1980s to find out what happens when humans are subjected to sensory deprivation, social isolation, and extended periods of confinement. Some of these experiments were funded by the US military, in response to alarming accounts of 'brain washing' techniques used in the Soviet show trials of the 1940s and by the Chinese government against Western prisoners in the Korean War. Further impetus came from NASA, whose interest in isolation studies as an analogue for long space missions continues to the present day. However, the Canadian government funded much of the most important research in the 1950s and 1960s, and a great deal of it was carried out at Montreal's McGill University.

One of the first of the Canadian studies, published in 1954, was by Bexton, Heron, and Scott. It remains one of the best-known academic papers on sensory deprivation, both for the remarkable results it revealed and the entertaining way in which those results were communicated.[7] The paper described the disturbing experiences of a group of male student volunteers who spent up to a week in solitary confinement. They wore translucent goggles, which meant they could tell the difference between light and dark but had no other visual input. The cubicle in which each individual was confined was soundproofed, and all they could hear was the background hum of the air conditioner. Cardboard cuffs, to restrict tactile stimulation, extended from the elbow and covered the hands, precluding any self-stimulation.

The volunteers tended to sleep a lot at the start of the study, but as time passed they lay awake, agitated and bored. In the absence of external sensory input, they generated their own noise by whistling, singing, or talking to themselves. They found it difficult to concentrate and unable to focus on simple cognitive tasks such as basic mental arithmetic. 'Blank periods' occurred, during which they had no thoughts. They became restless and found the whole experience exceedingly unpleasant. Most of the volunteers could not tolerate the conditions beyond three days and dropped out, despite generous financial inducements.

Several of the volunteers who did stay the course reported experiencing a variety of visual, aural, and tactile hallucinations or delusions. Some found these amusing, others irritating. Visual hallucinations ranged from simple dots and lines to complex images such as 'a row of little yellow men with black caps and their mouths open' and 'a procession of squirrels with sacks over their shoulders, marching purposely across a snow field'. Some heard voices or music, or felt objects hitting their skin.[8]

The McGill isolation studies captured the imaginations of psychologists and public alike. By the end of the 1970s, more than a thousand scholarly papers had been published on the topic. Yet many of the most startling results resisted replication in subsequent research. One review in the late 1960s suggested that fewer than one in five participants had actually experienced hallucinations.[9] The climate of opinion was also changing, as sensory deprivation experiments increasingly became associated in the public mind with stories of solitary confinement, torture, and CIA 'mind-control' experiments. Personal (and in some cases physical) attacks on researchers by political activists prompted many to leave the field.[10]

Whatever the limitations of the early isolation experiments, it is clear that people do sometimes have hallucinatory experiences under conditions of sensory deprivation. Moreover, those conditions do not have to be as artificial as in the laboratory experiments. Extreme environments that expose people to sensory (and particularly visual) monotony can have strange effects on perception, especially when the monotony is combined with stressors such as sleep deprivation and hunger. Perhaps surprisingly, hallucinations are not uncommon in everyday life either.

The pioneering psychologist William James wrote in 1890 that hallucinations were: 'a strictly sensational form of consciousness, as good and true a sensation as if there was a real object there. The object happens to be not there, that is all'.[11] As with so much of his work, James's assertion that hallucinations are genuine perceptual experiences was remarkably prescient. Recent neuroimaging studies have revealed that hallucinations activate the same brain areas as genuine perception, which is why they feel so vivid and realistic.[12]

The causes and neural mechanisms of hallucinations are still hotly debated. However, a wealth of evidence shows that hallucinations can occur in any sensory modality. They are most commonly visual or auditory, but people also hallucinate tastes, smells, or touch sensations, and somatic hallucinations can involve sensations of pain, heat, cold, dampness, or movement.

Hallucinations, especially auditory hallucinations, are a cardinal symptom of schizophrenia, and they can accompany other disorders including PTSD, bipolar disorder, and delirium tremens. They can also be induced by certain sorts of brain stimulation and by psychoactive drugs such as LSD, psilocybin, and opiates.[13] However, hallucinations are also common in people who are not suffering from any psychiatric disorder and have not taken psychoactive drugs.

Most people are understandably reluctant to admit that they have hallucinated, which makes empirical research problematic.[14] However, confidential surveys give some indication of how common they are. One European survey of 13,000 adults reported that nearly 40 per cent had experienced hallucinations, 6 per cent hallucinated once a month on average, and 2 per cent hallucinated weekly.[15] Other surveys have found that as many as 15 per cent of the population have experienced auditory hallucinations, mostly involving voices, but also abstract sound or music.[16]

Visual hallucinations are common among sight-impaired people, a condition known as Charles Bonnet Syndrome (CBS). Studies suggest that around 10 per cent of visually impaired elderly people have episodes of CBS, involving simple visual hallucinations such as patterns or shapes, or more complex visions of people, objects, landscapes, text, or even musical scores.[17]

Visual monotony is a known trigger for episodes of CBS. Hallucinations can also be triggered by stressors such as anxiety, hypoxia, exhaustion, dehydration, and sleep deprivation. It is therefore unsurprising that hallucinations are relatively common in extreme environments. Some of the most striking examples come from nineteenth- and early twentieth-century accounts of polar exploration. Historian Shane McCorristine has highlighted how many early descriptions of the polar landscape include quasi-supernatural descriptors such as 'malevolent', 'mysterious', or 'animate'. References to 'sensed presences', mirages, and optical illusions are common, lending some early accounts a distinctly spiritualist tone.[18]

Scientific studies of people in extreme environments have uncovered similar themes. One such study in the 1990s investigated hallucinatory experiences among a small sample of high-altitude climbers. None of the climbers had a history of head injury or of psychiatric or neurological disorders. Even so, they reported a variety of hallucinatory experiences, mostly involving the perception of their own bodies. Some had felt they were floating, or that their body had changed in size, shape, or weight. Others reported out-of-body ('autoscopic') experiences. They also experienced visual and auditory hallucinations, such as seeing people or animals, and hearing voices and bells.[19] These hallucinations seemed to be triggered by dangerous or otherwise stressful situations and were more common above 6,000 metres. One mountaineer, who was bivouacking solo on the edge of the 'death zone' at 7,500 metres, reported feeling his body becoming five times as large. Another, watching a fellow climber fall, said he felt himself 'projected from the rock' where he was standing and flying several metres through the air. Other research has similarly suggested that hallucinatory experiences are more likely to occur in life-threatening conditions.[20]

Many explorers and mountaineers have reported sensing the presence of an imaginary person.[21] This phenomenon, sometimes described as the 'sensed presence', occurs in people with neurological or psychiatric conditions such as migraine, epilepsy, depression, and schizophrenia.[22] But healthy individuals under great stress and in conditions of sensory deprivation have also described the presence of another person, often giving helpful advice. Numerous examples have been recorded in the personal accounts of people in hard places.[23]

The mountaineer Frank Smythe experienced a sensed presence in 1933, while struggling alone on Everest in brutal conditions. He kept feeling that another climber was with him, keeping him safe and offering comfort. So strong was this sense of companionship that at one point Smythe offered this 'ghost' a share of his Kendal mint cake. In the case of US astronaut Jerry Linenger, the imaginary companion was his father, a comforting presence during weeks of monotony on the Mir space station, which had been crippled by a series of near-fatal accidents. Linenger's father had been dead for many years. However, the astronaut regarded this as a manifestation of a psychological coping mechanism rather than anything supernatural.[24]

Various explanations for the phenomenon of the sensed presence have been put forward. For instance, it has been proposed that a sensed presence is a 'phantom body'—a variant of 'phantom limb syndrome', whereby individuals who have lost a limb nevertheless have the strong sense that it still exists.[25] Another theory posits that sensed presence experiences are caused by the influence of weak magnetic fields on the brains of susceptible people.[26] As with other forms of hallucination, however, the exact mechanisms remain unclear.[27]

Hallucinations occur in the absence of a real stimulus. Illusions, on the other hand, are mistaken perceptions of genuine physical stimuli. (It is possible that some extreme experiences that are reported as hallucinations are in fact illusions.) When we perceive something, our brain constructs a representation of that stimulus based on the sensory inputs. Normally, those sensory inputs are clear and it is easy to tell what we are seeing, hearing, or touching. But the ease with which we construct our representation of the external world lies on a continuum. The more ambiguous or faint the sensory stimuli are, the more assumptions the brain has to make in order to decide what reality is, and the harder it is to construct a representation. Eventually a threshold is reached, beyond which the brain decides that the stimulus does not exist, and therefore we perceive nothing. That threshold depends on several factors, including our expectations and the extent of other sensory input.

If we are expecting to see or hear something, our brain is quicker to construct the corresponding internal representation. And the more competing stimuli there are in the environment, the more obvious a particular stimulus must be before it is noticed. Conversely, in a monotonous environment even very faint stimuli may be noticed. When presented with very faint stimuli, the brain can find it difficult to decide what those stimuli represent, so it produces a 'best fit' response. Sometimes that initial 'best fit' does not make sense. If we are alert and thinking, we are likely to realize this and take a closer look. For example, psychiatrist Glin Bennet reported how one solo sailor saw an object in the water and initially thought: 'a funny place to put a baby elephant'. Forcing himself to concentrate, the sailor looked again and realized it was a whale.[28]

When someone is under stress, their cognition may be impaired and they are consequently less able to engage in complex reasoning about their perceptions. This may result in odd perceptions being accepted as real without further questioning.[29]

Before leaving this topic, we should note that the effects of sensory deprivation are by no means all bad. After experiments on extreme sensory deprivation fell out of favour, the psychologists who remained in the field followed a less controversial line of investigation. Their research suggested that sensory deprivation (by now renamed 'stimulus reduction') could have positive effects. This gave rise to the Restricted Environmental Stimulation Technique, or REST, popularly known as flotation tank therapy. Studies have indicated that REST can help in a range of conditions, including burnout, fatigue, and chronic pain. People employing REST have reported positive effects ranging from lower blood pressure and release of muscle tension to mild euphoria and even enhanced creativity.[30]

Having nothing to do

A less extreme form of monotony is simply having nothing to do to. Such tedium may be commonplace and unlikely to trigger hallucinations, but that does not mean it is trivial. The strain caused by having nothing to occupy one's time or attention for long periods can lead to lethargy and depression. It can also exacerbate social conflict within groups.

Most of us occasionally feel a bit bored. In some extreme environments, however, people may be exposed to long periods of unremitting tedium. Deep-sea divers, for example, may spend days or weeks confined in tiny decompression chambers. Even conventional dives using self-contained breathing equipment bring an element of monotony, as the diver is forced to spend long periods waiting at decompression stops. As we saw in Chapter 1, extreme dives to depths of 250 metres or more can require 10 hours or more of decompression on the way back to the surface. 'There is no way to jazz up decompression', wrote one diver, 'It is simply tedious'.[31]

Experienced divers find various ways of coping with the tedium of decompression. One renowned cave diver would enter a trancelike state, in which he would float motionless in a horizontal position for hours, apparently unaware of time.[32] Modern dive computers offer video games to entertain the decompressing diver. One diver we spoke to told us that some read magazines (glossy paper holds up remarkably well for a while under water) and pornography is popular in some circles.

Wilfred Thesiger faced tedium during his travels across the Arabian deserts. Describing one journey, he wrote: 'We rode for seven and eight and nine hours a day, without a stop, and it was dreary work. Conversation died with the passing hours and boredom mounted within me like a dull ache of pain'.[33] Thesiger found that the best antidote to the monotony of desert travel was to pay careful attention to the tiniest details: an insect under a bush, the tracks of a desert animal, or the shape and colour of the ripples on the sand. In this way, the very slowness of the journey became a remedy for boredom rather than its cause. Indeed, Thesiger mused on how rushing across the desert in a car would be far more boring than crossing it slowly on foot.

Having nothing to do at times is an occupational hazard for the military, including elite fighters. One US Special Forces operator recalled how he and his colleagues developed rituals for killing time during the long, empty interludes between missions, otherwise known as the 'hurry up and wait routine'. Some operators found relief by immersing themselves in the minutiae of grinding, brewing, and then savouring the perfect cup of coffee. Producing and drinking one cup could take them an hour.[34]

Mountaineers spend much of their time sitting around waiting for suitable weather. In one study, six climbers kept a diary of their activities during a forty-four-day expedition to climb the Thalay Sagar peak in the Western Himalayas. Only about 4 per cent of their total time involved actual climbing or travelling. The rest was consumed by mundane activities such as sleeping, eating, talking, 'thinking' (2.4 per cent), and doing 'nothing' (6.2 per cent).[35]

Even astronauts get bored. The Apollo 17 astronaut Gene Cernan wrote: 'Funny thing happened on the way to the Moon: not much. Should have brought some crossword puzzles'.[36] In a similar vein, US astronaut Norm Thagard wrote of how life on the Mir space station was mostly mundane. Aboard the International Space Station, highly trained astronauts must devote much of their time to the rote work of maintaining and running equipment. One of the essential attributes for any modern-day space explorer is the capacity to tolerate boredom and low levels of stimulation.[37]

To make matters worse, some of the people who are attracted to extreme environments are those who prefer plenty of novelty and sensory stimulation. Indeed, a strong aversion to boredom is a prime motivator for some individuals to seek out extreme experiences in the first place.[38] (We return to this theme in Chapter 12.) Ironically, having signed up for adventure and intense stimulation, they sometimes find that their biggest challenge is coping with long spells of inactivity and monotony.[39]

Tedium is capable of provoking reckless behaviour. Examples include an astronaut who left his capsule without fixing his safety tether, and experienced individuals from polar stations going outside in bad weather without the right clothing or equipment.[40] Dangers can also arise in less extreme environments. In 1978 an investigative reporter found that security personnel at a US Army depot were coping with the monotony of their work by sleeping, playing cards, racing their vehicles around the base, and getting drunk. Regrettably, the base in question happened to be one of the world's largest stores of nerve gas.[41]

Coping with tedium

The tedium of prolonged inactivity, environmental monotony, and even sensory deprivation may be unavoidable in some extreme environments, but practical measures can ameliorate their negative effects. For centuries, expedition leaders have recognized the pernicious effects of monotony and consequently the importance of keeping people stimulated and amused. Traditional remedies include talking, reading, eating, drinking, smoking, playing games, and writing, supplemented now by modern technology.[42]

Sleeping offers a partial solution, although there are limits to how long anyone can spend asleep. By the second winter of Adolphus Greely's Arctic expedition, his men had developed a habit of dozing in bed for up to 16 hours a day. Eventually, Greely prohibited bed rest between 8.00 a.m. and 3.00 p.m., except on Sundays. He nonetheless recognized the attraction of sleep and wrote that 'it was only by effort that I reduced my own sleeping-hours to nine daily'.[43]

The members of Shackleton's 1908 Antarctic expedition diverted themselves by playing music on a wind-up gramophone, reading books aloud to each other, and performing amateur theatricals. They whiled away hours in rambling arguments, known as 'cags', in which they debated theoretical issues such as why the wind blew in certain directions.[44] A piano sat in the cramped living quarters of Nansen's ship *Fram*, and fiddles, pipes, and guitars were essential inventory items. Many early polar teams put on plays and concerts, in which cross-dressing seems to have been a ubiquitous feature.[45] Even on the cramped International Space Station there is room for a bag of accessories with which to celebrate holidays, including party hats and a Christmas tree.[46]

Technology has given those in isolated extreme environments greater access to entertainment, prompting some authorities to worry about its suitability. Administrators once debated whether *The Thing*, a horror movie about an alien creature attacking an Antarctic station, should be allowed in Antarctic stations—only to find that it had been a popular choice in Antarctica for many years, with no apparent ill effects.[47]

One official encountered some distinctly odd behaviour when he arrived at an isolated Australian base on the sub-Antarctic Macquarie Island in 1950. The winter crew had a small collection of films, of which their favourite was the 1941 classic *Pride and Prejudice*. They had watched it repeatedly, to the point where they knew the dialogue by heart and recited it during showings with the volume turned down. Jane Austen's world had become their world, and they behaved in ways that were more in keeping with an early nineteenth-century manor house. At dinner, the act of passing the butter was likely to be greeted by the response: 'such affability, such graciousness—you overwhelm me'.[48]

The communal watching of movies can strengthen social bonds in confined environments, whereas private entertainment systems, such as internet-enabled computers and games consoles, can have the opposite effect. Some researchers have reported that when personnel in Antarctic stations acquired the option to watch videos in their private quarters, some of them became more withdrawn. Certain methods of passing the time are best enjoyed alone, however, as in the case of the Antarctic construction worker who deliberately overheated his room and, with the help of photographs and a vivid imagination, relived a family beach holiday.[49]

People in hard places sometimes turn to alcohol or other psychoactive drugs for relief from anxiety or boredom.[50] Crews in Antarctica, for instance, drink more and smoke more during the stressful and boring winter period compared to during the austral summer.[51] Tobacco, a mainstay of early polar explorers, has the dubious distinction of being one of the most highly addictive of all psychoactive drugs. (Nicotine is more addictive than barbiturates, amphetamine, cannabis, LSD, or ecstasy.[52])

During Shackleton's second Antarctic expedition, some of his men were stranded on the bleak and isolated Elephant Island for four and a half months. So strong was their craving for tobacco that they resorted to smoking penguin feathers, and one tried to smoke the wood from the bowl of his spare pipe. When Shackleton eventually returned to rescue them, after making an epic voyage in an open rowing boat, he threw bags of tobacco ashore before he even landed.[53]

One of the oldest and simplest ways of relieving tedium is by eating and drinking. Preparing and eating food can enhance psychological well-being in extreme or confined environments, and the cook is generally regarded as one of the most important people on a polar station or submarine. Meals help to mark the passage of time, which is important when every day is the same. Special occasions such as birthdays, Christmas, or national holidays may be marked with special meals, providing an excuse to indulge in modest pleasures.[54]

The eighteen-man crew of the *Belgica* were the first people to spend the winter in Antarctica, after their ship became trapped in ice during its voyage of exploration in 1898–1899. They entertained themselves with elaborate Sunday dinners, which helped to 'mark time and to divide, somewhat, the almost unceasing sameness' of their existence.[55] They took every opportunity to celebrate anniversaries, including the birthday of King Leopold, the legal holidays of all the nations represented, every team member's birthday, and Independence Day.

Communal dining often brings social benefits, especially for isolated teams. Indeed, some people may show up for meals primarily for the social contact, with the (perhaps inevitable) consequence of significant weight gain.[56] However, eating together is not always harmonious. The Ronne expedition of 1947–1948 disintegrated into warring factions by the end of their winter in Antarctica. Meals were eaten communally, but with two opposing camps at either end of the dining table and 'no man's land' in between. While one end of the table laughed and joked, the other end was quiet.[57]

Food and drink can bring other problems, ranging from pilfering, petty arguments, and jealousies to nutritional boredom and excessive consumption. Overeating can impair decision making and alertness. Even when food is plentiful, lack of variety can dull the appetite. The *Belgica* was loaded with fine food, including 'hashes under various catching names', 'sausage stuffs in deceptive forms', 'meat and fish balls said to contain cream', 'mysterious soups', and 'the latest inventions in condensed foods'. Yet the crew became more and more bored with the rich, mushy food, to the point where they would rather go hungry. The ship's doctor wrote that 'as a relief, we would have taken kindly to something containing pebbles or sand. How we longed to use our teeth!'[58]

The food in extreme environments is often downright horrible. During his desert treks, Wilfred Thesiger rarely had enough to eat, and the food he did have was often vile. Of one journey, he wrote: 'I had been hungry for weeks, and even when we had had flour I had had little inclination to eat the charred or sodden lumps which Musallim had cooked. I used to swallow my portion with even less satisfaction than that with which I eventually voided it'.[59]

The challenge of providing an attractive and palatable diet for people in extreme environments has been the subject of considerable research.[60] In 1969, NASA sent six men on a month-long research dive in the *Ben Franklin*, a submersible measuring just 48 feet by 10 feet, with the aim of studying how a crew functioned in an isolated confined environment analogous to a space station. One aspect that received detailed attention was their food preference. Each day the men recorded what they ate and what they thought of it. Although they accepted their pre-packed food, they did not particularly enjoy it: puddings were 'too sloppy', nut roll portions were 'too large', beef jerky was 'poor quality', and so on. The crew talked a lot about food, and post-exit interviews made it clear there was room for improvement.[61]

Food that is initially palatable can become less so as crews habituate to it. Studies dating back to the 1950s have shown that people find food less and less palatable the more frequently they consume it.[62] Having the ability to select your own meals seems to help. In a study conducted in an underwater habitat, participants who were able to choose their meals were happier than those who had to accept whatever they were given, even though the quality of food was the same in both cases.[63]

Food that is usually enjoyable can taste less pleasant when the environment affects people's sense of taste and smell. In zero gravity, for instance, the inability of the sinuses to drain gives astronauts the equivalent of a permanent head cold, and a consequent craving for spicy food.[64] Saturation divers, who live and work at depth for days or weeks at a time, also crave spices: the combination of breathing unusual gases and the effects of pressure on the food itself changes how it tastes. Bread, for instance, has 'the consistency and texture of a rubber ball'.[65]

The design of an isolated and confined environment can have a significant impact on the problems of monotony. When NASA built the Skylab space station in the early 1970s, they turned to the internationally renowned industrial designer Raymond Loewy for advice on habitability. Many of his suggestions have now become standard in confined environments.[66] Research on life in polar stations, submarine habitats, and space stations has found that varied colour schemes, moveable furniture, plants, and even the introduction of artificial smells can help to relieve the stress of monotony.[67]

One of Loewy's most influential suggestions—controversial at the time— was to include windows in Skylab to address the problem of visual monotony. Simply looking out of a window can relieve stress in someone living in a confined space.[68] One survey of astronauts found that the longer the mission, the more important it was for them to have the opportunity to gaze out of a window.[69] (On the Apollo Moon missions, the astronauts yearned to look down at the Moon's surface, but they had so much to do in preparation for their landings that there was little time to stop and stare.[70])

A similar phenomenon is found among submariners, who place a high value on 'periscope liberty'—the chance to remind themselves of the world outside their metal tube.[71] If no windows are available, films or pictures can take their place. Those that include scenes of natural settings have been particularly popular in confined environments such as submarines or polar stations.[72]

In conclusion, monotony and lack of environmental stimulation are common in extreme environments, as they are in everyday life. Extreme sensory deprivation can have adverse psychological effects, including hallucinations. Little pleasures such as eating, socializing, and playing games are often the simplest and best remedies for monotony.

6

Alone

I never found the companion that was so companionable as solitude.
Henry David Thoreau, *Walden* (1854)

The pursuit of certain extreme activities can involve periods of solitude ranging from a few days to many months. Whether solitude is a goal or an unwelcome consequence of their mission, solitary adventurers must contend with unique psychological challenges. The monotony of some extreme environments can do strange things to a person, as we saw in the previous chapter, and being alone makes it worse. Prolonged isolation can have detrimental effects on mental health. However, many individuals find solitude a positive experience, and for some the opportunity to be alone is one of the attractions of an extreme environment. In this chapter we examine the consequences of such isolation and consider why some individuals emerge psychologically intact, while others are broken.

Going solo

A desire for solitude is, for some individuals, the primary motivation for venturing into an extreme environment. Admiral Richard Byrd wrote that one reason he spent months alone during the Antarctic winter of 1934, with almost fatal consequences, was a 'desire to know that kind of experience to the full . . . to taste peace and quiet and solitude long enough to find out how good they really are'.[1] For most adventurers, however, solitude is a consequence rather than a cause of their adventure. Some cope well with being alone, and even come to enjoy it. Others do not.

Morton Moyes, a member of the 1911 Australasian Antarctic Expedition, spent ten weeks alone in a hut in the middle of the Antarctic after his companions went off to explore. Being isolated in the barren environment soon began to trouble him. In his diary Moyes wrote: 'The silence is so painful now that I have a continual singing in my left ear, much like a barrel organ, only it's the same tune all the time'. When his colleagues eventually returned, Moyes was speechless with emotion and so overjoyed that he stood on his head. Years later, he told a relative that it was not the loneliness as such that had oppressed him, but what he described as 'a sense of acute aloneness . . . surviving like the last leaf on a branch' in the desolate wilderness.[2]

Sailing solo around the world has been described as 'months of solitary confinement with hard labour'.[3] The first person to do so, at the end of the nineteenth century, was Joshua Slocum, whose account of his circumnavigation became an inspiration to later solo sailors. In it, Slocum vividly conveys the emotional highs and lows of his voyage. Although initially he felt great loneliness, Slocum wrote that once this had passed 'the acute pain of solitude' did not return.[4] He finished his voyage apparently unscathed, but later suffered deteriorating mental health and found it difficult to settle. Ten years later, Slocum sailed for the West Indies. He was never heard of again.

Technology has allowed more recent solo adventurers to remain in communication with the outside world. Even so, the psychological pain of solitude can add greatly to the physical hardship.

Choosing solitude

Prolonged solitude may be relatively unusual in everyday life, but most of us do spend time alone, whether by choice or circumstance. Many people do not particularly enjoy being on their own, or enjoy it only in small doses, but some actively prefer it.

A preference for solitude can develop for a number of distinctly different reasons. In some cases, individuals who prefer solitude may have psychological problems. Those who are socially anxious, or have limited social skills, may avoid social situations because they find them stressful and unpleasant. Depression often leads people to withdraw from social contact, although this can reflect apathy about activities in general as well as an aversion to social contact.[5]

For a small minority, the avoidance of social contact may be a manifestation of an extreme and enduring personality type. Such individuals are genuinely uninterested in other people and get little or no pleasure from social contact. Social anhedonia, as this trait is known, is a cardinal symptom of a cluster of personality types known as schizoid and schizotypal personalities. Individuals who display schizoid patterns of behaviour and thinking are typically detached and withdrawn, showing little interest in intimate relationships, friendships, or other people in general. Schizotypal individuals also tend to hold odd or eccentric beliefs and are more likely to experience hallucinations.[6]

Social contact helps to protect us against a range of psychological problems, which means that those who avoid it lose out. One longitudinal study found that individuals who showed signs of schizotypy in adolescence displayed more extreme symptoms in their late twenties. The researchers suggested that when these individuals were adolescents, they found it difficult to withdraw from social contact as much as they might have preferred because of restrictions placed on them by parents and school. However, once they were older and freer to make their own choices, they chose social withdrawal, thereby losing the psychological benefits that social interactions afford.[7]

By no means everyone who is attracted to solitude has a mental health problem. Many individuals who prefer their own company are psychologically healthy. They have variously been described as 'happily introverted' and 'unsociable' (as opposed to anti-social).[8] They choose to spend time alone, find it enjoyable, and show no signs of mental illness. On the contrary, they report high levels of happiness, good social relationships, and a striking lack of loneliness, all of which suggests that solitude can contribute to mental health. These 'happy introverts' have been somewhat overlooked by research, which has tended to focus on the negative consequences of solitude.

Psychologists have coined the term solitropism to denote a desire to spend time alone, which they contrast with sociotropism, meaning a desire to spend time with other people. A desire for solitude, they suggest, could be a result of either high solitropism or low sociotropism. In other words, people who choose to be alone may be either solitude seekers or misanthropes. Research suggests that people who report enjoying solitary activities do so because of a strong desire for solitude rather than a weak desire to socialize.[9]

One way of measuring the tendency to prefer one's own company is by using a psychological tool developed by psychologist Jerry Burger, known as the Preference for Solitude Scale (PSS). Burger recognized that we can enjoy being alone some of the time and yet value being with other people. He therefore distinguished between individuals who seek solitude because they are socially anxious and those who seek solitude because they positively enjoy spending time alone.[10]

Burger's research found, unsurprisingly, that individuals who scored highly on the PSS spent more time by themselves and reported enjoying it more. The diaries they kept as part of the research confirmed as much. However, Burger was surprised to find a strong correlation between a preference for solitude and loneliness. The two concepts are usually contrasted, because preferring to spend time alone seems distinctly different from feeling lonely. However, Burger suggested that the two might be more compatible than is often supposed. You can enjoy being alone yet still have a desire for more, or perhaps better, social relationships.

Subsequent research has shed more light on this seemingly paradoxical link between preference for solitude and loneliness. One study found that the overall PSS score was made up of three distinct, and sometimes conflicting, elements: the person's need for solitude, the degree to which they enjoy being alone, and their perception of their productivity during solitude. High PSS scores were found to correlate with loneliness only among those individuals who scored highly on their need for solitude and their enjoyment of solitude. In contrast, people who were productive during solitude scored lower for loneliness.[11] Other research found that spending time alone through choice could result in three distinct types of outcome: feeling more effective at problem solving and creativity, a sense of intimacy or spirituality, or loneliness (the only clearly negative outcome).[12]

Taken together, these studies suggest that the people who cope best with being alone will be those happy introverts who value the opportunities that solitude offers for creativity, productivity, or spirituality, rather than those who simply want to be on their own.

How does someone develop into a happy introvert? The psychiatrist Anthony Storr argued that the capacity to enjoy solitude has its roots in childhood and is a learned trait.[13] The evidence is sparse, however. Research on children who enjoy solitude has been much less extensive than research on shyness, social anxiety, and ostracism. Parents and teachers often find it hard to understand why a child withdraws from social activity, and whether such withdrawal is harmful or benign.

Psychologists have used the term unsociability to refer to a child's preference for solitude and solitary activities, contrasting this with shyness or ostracism by peers. Unsociable children are those who seem content to play alone but are willing to engage in social activities if invited. They do not differ significantly from others in their social abilities: once they become engaged in conversations, for instance, they participate as much as their more sociable peers. Moreover, there is little evidence that they are lonelier or more likely to avoid school. Indeed, one study found that unsociable children reported liking school more than did comparison children.[14]

The developmental origins of unsociability are not clear. One study suggested that children who were unsociable tended to have parents who place less emphasis on the importance of making friends, but this is likely to be only one element in the unsociable child's development.[15]

The consequences of solitude

People who are psychologically healthy and have a preference for being alone may be well equipped to cope with protracted solitude. But what happens to those who are not so well placed? The negative consequences of solitude range from occasional feelings of loneliness to psychological breakdown. Let us start with loneliness—that sad yearning for the company of others.

A person does not have to be alone to feel lonely. One can be surrounded by people and yet still feel intensely lonely. Wilfred Thesiger was never lonely while exploring the Arabian deserts, but he experienced great loneliness at school and in European towns where he knew nobody. 'The worst loneliness', wrote Thesiger, 'is to be lonely in a crowd'.[16]

Loneliness is a subjective state—it depends on your perception of the quantity and quality of social support you receive. Our sense of what is the 'right' amount of social support is idiosyncratic. One person may be happy having emotionally intimate contact with just one or two close friends, while another person may have close friends and still feel lonely unless they also enjoy frequent contact with numerous acquaintances.[17] Each of us appears to have a subjective threshold for social contact, below which we start to feel lonely. This individual variation helps to account for the evidence that lonely people do not necessarily spend more time alone than people who do not regard themselves as lonely.[18]

We all occasionally experience feelings of loneliness—for instance, when moving to a new town or after a relationship breaks up.[19] Some people, however, develop a persistent and deep-seated sense of not belonging. This state of enduring loneliness can have long-term negative consequences for physical and psychological health. Extensive evidence shows that persistent loneliness takes a physical toll in the form of impaired cardiovascular function and less effective immune function. Indeed, some experts have compared the health impact of loneliness to that of obesity or smoking. Lonely people are also less resilient to stress. Social support is one of the best buffers against stress, and its absence leaves people more vulnerable to the physical and psychological effects of stress that we described in Chapter 1.[20]

What is more, loneliness is often self-reinforcing. Persistently lonely people tend to be more socially anxious, have lower self-esteem, perceive slights and rejections when none were meant, and trust others less. Unsurprisingly, this pattern of negative thoughts and behaviour is associated with greater difficulty in forming relationships.[21] Breaking out of this vicious circle can be very hard.

In extreme environments, persistent feelings of loneliness can easily develop into depression. Richard Byrd's diary shows that after several weeks alone in his Antarctic hut he developed a form of apathy that is typical of depression. 'Why not let things drift', he wrote, 'in the direction of uninterrupted disintegration . . . everlasting peace. So why resist?'[22] For individuals who are already vulnerable to mental illness, prolonged solitude can be highly detrimental.[23]

This appeared to be the case for the climber Johnny Waterman, who was last seen in 1981 heading up Alaska's Mount Denali. He came from a troubled family of climbers.[24] His brother Bill had disappeared seven years previously, after descending into depression and self-harm. His father Guy, who turned to climbing after battling alcoholism, suffered from depressive illness and craved the peace of the wilderness.

In 1978 Johnny spent several months alone on Alaska's Mount Hunter, an experience that hastened his downward spiral into a form of mental illness that he dubbed 'Mount Hunter psychosis'. He eventually checked himself into a psychiatric hospital, where he was diagnosed with schizoaffective disorder. After discharging himself from the hospital he headed for Mount Denali, where he attempted a winter climb, unroped and ill prepared. His death was listed as an accident but many believe he was suicidal. Two decades later, Guy climbed to the top of a New Hampshire mountain, lay down, and died from the effects of exposure.

Studies of solitary confinement in prisons have provided compelling evidence that prolonged solitude can seriously damage individuals who are vulnerable to mental illness. The evidence overwhelmingly shows that solitary confinement can have a severe, detrimental, and often long-term impact on psychological health. For prisoners who have a pre-existing mental illness, or a vulnerability to mental illness, breakdown is even more likely.[25]

Solitary confinement involves sensory deprivation and loss of control as well as social isolation, making it difficult to unpick the specific effects of the latter. However, a lack of interpersonal contact does appear to be important in triggering psychosis. When conditions are ambiguous or uncertain, we may rely on other people as a reference point for assessing reality. If you are unsure whether, say, a peculiar sound you heard was benign or threatening, checking with others can help you decide. The sustained lack of opportunity to sense-check one's own perceptions can eventually render someone unable to distinguish the objective, external reality from their own subjective, internal perception. Psychologists have described this phenomenon as 'confinement psychosis'.[26]

The prolonged solitude of extreme activities can have a catastrophic effect on the most vulnerable individuals. Donald Crowhurst appears to have been one such person. In 1968 Crowhurst set sail on the Golden Globe yacht race—a non-stop, solo round-the-world race described as 'a voyage for madmen'.[27] The ebullient British entrepreneur had neither a seaworthy boat nor the skills to tackle such a journey. But he had staked his home, his business, and his pride on the adventure. Despite last minute doubts, he could not bring himself to walk away.[28]

One by one, the Golden Globe racers dropped out, some because of storm damage, others because they could no longer endure the painful loneliness. Just one man finished. Another sailor found the experience of solo sailing so addictive that he could not stop: he abandoned the race but carried on sailing, eventually circumnavigating twice. For much of the race, Crowhurst appeared to make remarkable progress, but his readings were fabricated. In reality, he never left the Atlantic, where, eight months after his departure, his empty boat was found drifting.[29]

During his solo voyage, Crowhurst progressively lost touch with reality. His state of mind can be inferred from the logs he kept. These gradually degenerated from a lucid factual account into grandiose quasi-philosophical ramblings, which some psychiatrists later judged to be symptomatic of psychosis. As it became clear to Crowhurst that he could not finish the race, and that his attempts to fabricate evidence would be uncovered, his hopes disintegrated. Faced with the prospect of crippling debts and public humiliation, he withdrew into a fantasy world in which he was the supremely insightful master who had unlocked the secrets of the universe and existed on a par with God. He is believed to have committed suicide.[30]

In addition to the psychological challenges, being alone in extreme environments heightens the physical risks. When there is no one else to advise, assist, or rescue you, risky exploits become even riskier. In 2001, Canadian Robert Kull spent a year alone (except for a cat) on an isolated island off Chile, studying the psychological effects of solitude. He sustained several injuries caused by sliding on the rocks, not helped by the fact that he had a prosthetic leg. After three months of repeatedly injuring himself, he wrote: 'I'm beginning to see the rock out front as actively malevolent—watching and waiting for the chance to harm me'.[31]

Despite its hazards, solitude and the freedom it can afford may be a positive experience. There is ample evidence that being alone, at least for limited periods, can be good for us—for example, by stimulating creativity or helping to reach important decisions.[32] Throughout history, people have chosen to withdraw briefly from society in order to think through problems or develop new perspectives.

Even long periods of solitude can be psychologically healthy, under the right circumstances. As part of her search for silence, the author Sara Maitland spent six weeks alone in a remote cottage on the Scottish island of Skye. She later wrote a beautiful book about her own and others' experiences of solitude, highlighting its many positive aspects. Maitland found that her senses became sharper and sensations more intense. She wrote, for example, of how porridge 'tasted more like porridge than I could have imagined porridge could taste'.[33]

In the absence of other people, the solitary adventurer sometimes forms a relationship with their environment. During her retreat, Maitland's sense of herself and her surroundings began to change. She felt increasingly disinhibited: she could do whatever she wanted and there was no one to judge her. But she also became acutely aware of her place in her environment, to the point where she felt a 'sense of vast connectedness, of oneness with everything' and entered a 'state of bliss—a fierce joy, far beyond happiness or pleasure'.[34]

A feeling of connectedness with the environment is not unusual. Many people who have entered extreme environments have described a sense of unity with the world, of feeling they were no longer a discrete individual. Solo sailors have written of moments when they felt they were part of their boat, or of the ocean or sky. A polar sojourner wrote of merging with the moonlight in the mountains, while a mountaineer felt 'an extremely intimate oneness with the universe'.[35]

Even brief periods of isolation, when combined with an extreme activity, can trigger such feelings. In 1957 one of the pioneers of US military high-altitude balloon experiments spent 32 hours at an altitude of up to 100,000 feet in a pressurized capsule suspended below a huge helium balloon. Part way through his journey he felt as though he 'no longer belonged to the earth' and that his identity was 'with the darkness above'.[36]

Solo pilots sometimes experience a phenomenon they call 'break-off', which has been described as feeling 'isolated, detached, or separated physically from the earth'. A study in the 1950s found that around a third of high-altitude pilots interviewed by US Navy researchers had experienced break-off, particularly when they were at high altitude, without radio contact, and with not much else going on. 'You get break-off when you have a chance to be free about your thoughts', said one. Most described it as a 'very personal', even spiritual, experience that was not the sort of thing flyers talked about. Some pilots, however, did not enjoy the sensation, and felt anxious until they had descended closer to the ground. More than two-thirds of those who experienced break-off described it as lonely.[37]

Solitude can leave individuals feeling closer to their own self, or to some divine or spiritual presence. Divine revelation during solitude is a feature of most major religions, and followers may spend time alone in 'retreat' in order to come closer to their god. Others who have not set out with spirituality in mind may nevertheless feel that spending time alone has altered their sense of self.

Coping with solitude

How might someone prepare, both practically and psychologically, to cope with solitude? Many explorers have thought about what is required.

Augustine Courtauld spent five winter months alone in an Arctic weather station in the 1930s. In his account, Courtauld argued that such isolation was not necessarily damaging, as long as the individual had volunteered for the task and was sure they would eventually be relieved. They should have 'an active, imaginative mind, but not be of a nervous disposition'[38] and have plenty to occupy their thinking. Courtauld passed the time by drawing plans of boats, shovelling snow, reading, and maintaining his meteorological instruments.

Courtauld's assertions about the ideal personality for isolation remain speculative. The sensory deprivation research that we discussed in the previous chapter found considerable individual variation in people's ability to cope with prolonged isolation and sensory deprivation. But despite extensive analysis, no clear associations with personality were found.[39]

Having some prior experience of isolation can help. Nicolette Milnes Walker, a psychologist who was the first woman to sail single-handed across the Atlantic, attributed her ability to cope with isolation to her experience in boarding school, which had given her a 'habit of solitude'. The fact that she had chosen her situation also helped: 'I planned it, I expected it, and I expected it to end'.[40]

Anecdotal accounts such as these are supported by evidence from isolation experiments. The participants in these experiments generally found that isolation became easier to bear the more often they experienced it.[41] Similarly, a 1960s study of Norwegian children from isolated farms found that they coped better in 5-hour perceptual isolation experiments than did children from urban environments. The children who were familiar with being alone experienced fewer perceptual, cognitive, and emotional disturbances during the experiments than those who were used to a more stimulating environment.[42]

Although one might expect people who engage in solo expeditions to feel lonely, their accounts suggest that deep and persistent loneliness is the exception rather than the norm. They do experience transient feelings of loneliness, but rarely does loneliness appear to have become all-consuming.

(That said, the writers of these personal accounts might not always present a fully objective picture of their feelings.)

When feelings of loneliness do arise they are often ascribed to contextual factors. The sailor Robin Knox-Johnston felt lonely at Christmas—the first time on his solo circumnavigation when he felt he was missing an important social occasion. A couple of glasses of whisky and some boisterous carol singing helped to dispel the mood, as did completing the rituals of Christmas Day, including a toast to the Queen.[43]

Keeping busy is a good coping strategy. But even with plenty to do, the absence of social contact can still be hard to bear. The psychological need for affiliation is deep-seated, and solo adventurers tend to make the most of any opportunity for companionship, whether virtual or real.[44] While some contemplate their relationship with their environment, themselves, or their god, others form more prosaic connections.

Listening to talk shows on her short-wave radio gave Milnes Walker a sense of connection to the outside world. Naomi James, who circumnavigated the world solo in 1972, took a kitten to keep her company. Joshua Slocum tried to alleviate his loneliness by barking orders to imaginary crewmembers, but found that this reminded him there was no one there. Knox-Johnston learned poems and recited them to passing seabirds.[45]

After three months alone in his cave, Michel Siffre spent a week trying to catch a mouse to keep him company. Ironically, Siffre had previously exterminated the resident mouse colony because their noise irritated him. But he became so starved of company that catching the lone survivor grew into an obsession. He finally succeeded in tempting the mouse under a dish, only to kill it by mistake when he brought the dish down on the animal's head. Siffre was stricken with grief.[46]

Having other reasons for wanting to be in an isolated environment can also help to cope with the solitude. When researchers were recruiting volunteers for cave-dwelling experiments like Siffre's, they found that potholers, who had a pre-existing interest in caves, could stay longer because of their motivation. Geoffrey Workman, a potholer from Yorkshire, was highly motivated to spend time underground for another reason. In the early

1960s, when nuclear war seemed likely, there was great interest in how people might survive the apocalypse. Workman wanted to prove it was possible for humans to stay underground for long periods. He broke Siffre's record by spending 105 days alone in Stump Cross Cavern.[47]

We have seen, then, that solitude can do strange things to people. But many cope well with it and do not suffer psychological damage. Loneliness, on the other hand, is generally bad for mental and physical health. For most of us, periods of voluntary solitude can be a positive experience if they are approached in the right way—for example, as an opportunity for reflection or creativity.

7

Other People

Hell is other people.   Jean-Paul Sartre, *Huis Clos* (1944)

B eing alone, as we saw in the previous chapter, can be good or bad for
psychological health, depending on the individual and their circumstances. The same is true of being with other people. Company may bring
both companionship and conflict.

Other people can be life saving in extreme environments. As well as
providing practical help when it is needed, interactions with others relieve
monotony and camaraderie is a powerful motivator. But the presence of
other people is also one of the most testing features of extreme environments. Being confined for long periods with one person or a few other
individuals—in, for example, a cramped space station—can be highly stressful. As cosmonaut Valeriy Ryumin put it: 'all the conditions necessary for
murder are met if you shut two men in a cabin measuring 18 feet by 20 and
leave them together for two months'.[1]

For Wilfred Thesiger, the mental strain of living together with his small
group of Arab travelling companions was the toughest aspect of desert
travel—worse even than the physical strain of the heat, hunger, thirst, and
constant risk of attack. He wrote of how hard it was to live 'crowded
together with people of another faith, speech and culture in the solitude of the
desert' and how easy it was to be provoked to anger by their behaviour.[2]
Thesiger was conscious of how, under these extreme conditions, he could
easily form a violent dislike of another member of the party and use him as a
scapegoat. He was equally conscious of how utterly dependent he was on his

companions, who could at any time have murdered him, buried his body in the sand and made off with his possessions. Yet he trusted them completely and never worried about being betrayed.

In this chapter we consider what science has to say about interpersonal relationships in extreme conditions. We explore the sources of friction between team members and what happens when the pressure gets too much. We explain how bonds between people are built and strengthened, and what can happen when romance or sexual desires flourish. The evidence leaves little doubt that strong interpersonal skills are crucial to the success of any extreme mission involving more than one person.

Conflict

If you find yourself trapped with the same small group of people for a long time, the ability to tolerate others' foibles is an essential attribute. So is being tolerable oneself. In everyday life, most of us spend at least some of our waking hours with other people: at work, as part of a couple or a family, or with friends. But even the best of friends can fall out, and even a couple who are deeply in love can sometimes find each other's habits irritating. We may cope with these tensions by removing ourselves from the problem or mixing with different people for a while. If the friction becomes intolerable we can, if necessary, end the relationship or change job. These options are rarely available in extreme environments.

Personal relationships are much harder to manage in extreme environments, where there are few, if any, options to escape. Everyone is exposed to environmental stressors that interfere with concentration, disrupt sleep, and raise anxiety, all of which tend to foster social tensions.[3] Small quirks and personal habits start to grate. The polar explorer Richard Byrd referred to the moment when one has nothing left to reveal to the other, when even his unformed thoughts can be anticipated, his pet ideas become a meaningless drool, and the way he blows out a pressure lamp or drops his boots on the floor or eats his food becomes a rasping annoyance.[4]

When personal relationships break down, the result can be conflict and the formation of cliques. At worst, this leads to scapegoating, ostracism, and even violence, sometimes with dramatic consequences. In 1959, a Soviet scientist in an Antarctic research station killed a fellow scientist with an ice axe after an argument over a chess game. (The Soviet response was to ban chess at their Antarctic stations.)[5]

The twelve scientists who volunteered for the six-month International Biomedical Expedition to the Antarctic in the 1980s had a particularly bad time.[6] During the expedition they conducted numerous physiological and psychological experiments on themselves. These including taking cold baths to see if that would hasten acclimatization (it does not), being injected with toxins, monitoring sleep patterns with electrodes attached to their heads and bodies, and standing motionless in temperatures of minus 25 degrees Celcius to find out how long it took to freeze. They suffered the indignity of having their food, faeces, and urine measured daily. To cap it all, they had to cope with a TV crew that was along to make a documentary.

Understandably, the men did not enjoy the experience. Being injected with noradrenaline, for instance, left them in shock, and one was reduced to tears by the agony. They had inadequate rest and were tormented by the knowledge that another scientific expedition nearby was enjoying warm surroundings, top-notch food and wine, and regular radio contact with the outside world.

Yet it was not the physical strains that were most difficult to tolerate, but the strain of being with each other. The researchers who planned the expedition realized that they had underestimated the importance of compatibility for people immersed in a stressful environment. The expedition members came from five different countries and were divided by language and culture. One (unnamed) national subgroup was perceived by the others as hypercritical and 'imperialistic'.[7] The tensions were so bad for one scientist that he was evacuated after becoming withdrawn and depressed. The animosity festered for several years after.

Both the interactions between individuals and the context in which those interactions occur will help to determine whether relationships are harmonious. One of the most important contextual factors is lack of privacy, to which we return shortly. The avoidance of conflict also depends on picking the right team. This means not just selecting individuals with the right skills, knowledge, and personality characteristics (issues we discuss in Chapters 8 and 9), but also taking account of cultural factors that affect people's ability to get along.

Culture has been defined in many different ways. For our purposes, culture is the set of values, beliefs, and behavioural norms shared by a group of people that helps them understand and navigate the world, and which serves as a basis for social identity.[8] Differences in the way people behave, think, speak, and view the world are obvious potential sources of friction, and this often becomes more evident when those differences are attributed to national origin.

Most present-day space missions are international collaborations, and space agencies devote a great deal of effort to ensuring astronauts from many nations work harmoniously together. But this was not always the case. Before the International Space Station was established, the Soviets hosted several international guests aboard their space station Mir for periods of up to six months. Psychologists who studied these and other multinational space crews found that being a 'guest' astronaut on another nation's space mission could be a lonely, frustrating experience.[9]

One source of frustration was being prevented from carrying out meaningful work. For instance, foreigners were forbidden to touch any controls, with one joking that he had 'red hand syndrome' from being slapped every time he reached out. US astronaut Norm Thagard had to watch his Soviet crewmates doing all the interesting work, while his compatriot David Wolf ended up with the unglamorous role of 'wall cleaner' on board Mir. When American Shannon Lucid was left alone in Mir while her Russian crewmates went space walking, she found the control switches had been taped down.[10]

Relationships between minority guests and their hosts were generally friendly on shorter missions. However, mistrust and negative feelings tended to increase over the course of longer missions. Most of those studied said they would prefer to be with people of their own nationality on long space missions.[11]

Culture-related friction came to the fore in an isolation experiment involving three international crews, which ran from July 1999 to April 2000.[12] The aim of the study was to see how people coped on long-duration space missions by locking twelve people in a specially constructed mock-up of a space station. One crew was entirely Russian, but the other two crews included a mix of international participants. Different crews spent periods in isolation from each other and cohabiting. Hatches and tunnels connected the modules in the simulated space station, allowing ground control to isolate or mix the groups. At least, that was the plan.

What started as a sober scientific experiment garnered international media attention following two incidents. The all-Russian Crew 1 had got on well with Crew 2, which consisted of three Russians and one German. But tensions emerged when the truly international Crew 3 arrived, consisting of Russian, Austrian, and Japanese men, and a Canadian woman (the only female participant). Matters came to a head during a New Year party, when two Russians got into a bloody fight and another tried to kiss the Canadian. The alcohol did not help.

The members of Crew 3 were upset and blocked the connecting hatch to prevent further mixing with Crew 1. The Japanese astronaut left the study early. The remaining Crew 3 members then proceeded to fall out with each other. The sole Russian in Crew 3 was married to a member of ground control; the others became suspicious of the time he spent talking with her and ostracized him. They simmered in resentment at being ordered around by 'faceless foreign persons (who often spoke only Russian)'.[13]

Once the dust had settled, researchers tried to assess what had gone wrong. One of the main issues was cultural difference, compounded by the fact that most of the participants did not share a common language and had trouble communicating. The Russians thought Crew 3 had overreacted to incidents that would be seen as normal in their culture. Crewmembers did not know each other before the experiment and received no training or advice on cross-cultural understanding. They had been selected for their technical expertise and enthusiasm rather than social compatibility.

Although schisms often occur along national lines, groups may also fracture along professional lines. There is a long history of conflict between scientists and armed forces personnel on extreme missions. In the early days of space travel, astronauts sometimes resented sharing missions with non-military personnel. One Shuttle veteran referred disparagingly to the visiting specialists as 'part-time astronauts'.[14]

In the eighteenth century, the botanist and zoologist George Steller clashed with Captain Commander Vitus Bering during the Russian Kamchatka expedition. Steller was certain of his superior intellect, which he believed was divinely inspired. As the ship ventured into uncharted territory, Steller used his observations of marine flora and fauna to predict—confidently, but not always accurately—where land was to be found. Bering ignored him. Steller felt slighted, grumbling in his journal about the officers who 'mocked whatever was said by anyone not a seaman, as if with the rules of navigation all science and powers of reasoning were spontaneously acquired'.[15]

An anthropologist aboard an oceanographic research vessel in the 1970s reported similar conflict between scientists and the rest of the crew. It came to a head when crewmembers threw biological specimens overboard to make room in the fridge for alcoholic drinks.[16] More recently, the simmering tensions between scientists and workmen during winter-over missions in Antarctica led one observer to conclude that 'boffins and builders' do not get on.[17]

Even relatively homogeneous groups may find reasons to divide into cliques. Despite their common profession, backgrounds, and nationality, the thirty-three miners who were trapped underground for more than two months in the Chilean mine disaster of 2010 ended up dividing along family lines into 'sub-clans'.[18] Schisms can also occur within teams because of perceptions that some individuals are less sociable or are not working as hard as others.[19]

Ostracism

One of the worst forms of social conflict is when a single individual is ostracized by the rest of the group. Ostracism is a common phenomenon, prevalent throughout history and geography. It is not even unique to humans: ostracism has been observed in many other social species.[20]

Extreme environments have provided plenty of case histories. For example, a 1960 Antarctic expedition featured instances of ostracism of the cook, the biologist, and the signalman during the course of a year-long mission involving only fourteen men.[21] During another Antarctic expedition, to an Australian base on Casey Island in the 1980s, the officer in charge (OIC) was shunned by everyone and took to eating alone in a corner. A decade later, another OIC on Casey Island was ostracized by his crew, who had become weary of his adherence to the rules and habit of pinning up notices. He remained in charge, but with no authority: when he sat at the dining table, everyone would leave.[22]

Ostracism is felt most keenly when the victim sees the rest of the group continuing to socialize. This has been referred to as social ostracism, in contrast with physical ostracism when someone is physically excluded from the group. A person who is shunned while remaining physically part of the group may find it emotionally more painful than someone who has been banished.[23]

Individuals who are ostracized naturally tend to search for reasons. These may not be obvious, because the people shunning them may not explain why. Typically, those who are ostracized start to wonder if they have done something wrong, or if there is something wrong with them. They may even begin to feel like a 'non-person' who is unworthy of attention. Little wonder that being ostracized produces feelings of meaninglessness and low self-esteem.[24]

Research shows that it does not take much to make someone feel shunned. Even in small doses, social exclusion can lead to negative feelings and measurable reductions in self-esteem. Experiments involving more than 5,000 participants found that it took only 2–3 minutes of ostracism for negative feelings to emerge.[25] Similar responses occur in virtual environments as well.[26] You do not have to like, or even know, the people who are shunning you to feel the pain of exclusion.[27]

Individuals who are socially ostracized tend to display a common sequence of behaviour. Research shows that most people, when ignored, initially feel a sharp stab of emotional pain. (Incidentally, 'pain' may be more than a metaphor in this context: ostracism produces changes in brain activity similar in some respects to those observed when people experience physical pain.[28]) If the ostracism persists, they respond by attempting to re-establish relationships with their group. If they fail to get a response, they become frustrated, angry, or sad at their inability to influence the situation. They may step up their efforts to reintegrate, using ingratiation and conforming to group norms. This strategy sometimes backfires if the individual is then perceived as irritating.

Research further suggests that excluded individuals are more likely to engage in impulsive acts, and that ostracism makes people think less intelligently. It seems that the shock of rejection disrupts our ability to think clearly and regulate our own behaviour.[29] Some ostracized individuals respond with resignation, others with defiance. In extreme cases, ostracism has been linked to violence and some researchers have suggested that it may be a motivating factor in killing sprees. One study found that in thirteen out of fifteen school shootings in the US, the killers had previously experienced chronic social rejection.[30]

Almost everyone experiences the immediate 'pain' reaction to being ostracized.[31] Individuals differ, however, in their responses over the longer term. Women are found to be more likely, on average, than men to step up their efforts to fit in with the group—for example, by working harder in group tasks.[32] Individuals who are naturally socially anxious, and who score highly on the personality trait of neuroticism, are more likely to become withdrawn, whereas those who are quick-tempered may be more inclined to lash out.[33] Those who have a history of repeated rejection may develop 'rejection sensitivity': actively looking for signs that they are being shunned, bristling with hostility when they think it is happening, and avoiding situations where they might experience it.[34]

Those who endure prolonged ostracism can end up suffering from insomnia, agitation, and depression. In extreme cases, they may give up and lapse into helpless despair. Some may even lose touch with reality, experiencing hallucinations and delusions.[35] Clearly, it is not good for any team in an extreme environment to have someone in their midst who becomes violent when ignored. But neither is it good to have a team member who has given up.

Privacy

The physical environment plays a crucial role in fostering or easing conflict. Two important and related factors are personal territory and privacy.

In common with many other animals, humans have a propensity for territoriality. We tend to feel more comfortable and in control when we are in a space that we regard as our own, whether that is our home or a desk within an open-plan office. We may feel violated if that space is invaded without permission. We also tend to respect and avoid space that appears to have been claimed by someone else. Personal objects serve as markers of ownership and defensive barriers to others. When there are no natural opportunities to delineate personal territory, we tend to create our own.[36] In extreme and confined environments this can lead some individuals to occupy and defend space that others perceive as communal, an inevitable source of tension.[37]

Closely related to issues of personal territory are those of privacy. Our sense of whether we lack privacy depends on the discrepancy between what we subjectively regard as an optimal level of privacy and our actual level at the time.[38] Similarly, our perception of whether an environment is crowded is a subjective judgement, affected by the nature of the environment, our personal preferences, and our cultural background. For example, we are more likely to tolerate crowding in a public place than at home, and are more tolerant if we are used to it.[39]

Wilfred Thesiger found the lack of privacy to be one of the most stressful aspects of travelling in the desert—far worse than the physical discomfort. Every word he said was overheard and every movement he made was watched, for weeks or months on end. It was almost impossible for him to have a private conversation with one of his travelling companions. Even if they moved apart from the group, the others would come to listen and join in the conversation.[40]

Having a private space—somewhere to retreat from others—seems to be a universal human need, even if the circumstances in which we feel it necessary to withdraw vary from individual to individual (as we saw in the previous chapter). As Charles Darwin put it, when embarking on his voyage aboard the cramped *Beagle*, 'the absolute want of room is an evil that nothing can surmount'.[41]

Historically, people in hard places have always found imaginative ways of creating some form of privacy, whether behind a curtain in a bunk, in the corner of a ship's library, or by immersing themselves in their work—as did the ship's engineer aboard the Norwegian polar vessel *Fram*, who spent time alone in the engine room dismantling and rebuilding the engine.[42] The designers of space ships, polar stations, and other artificial habitats aim to incorporate opportunities for individuals to have at least a small space they can call their own. Ideally, these private spaces should have visual and acoustic shielding, so that individuals can spend time alone and out of sight of each other and, where relevant, ground control.[43]

Sex in extremes

Sex is usually missing from written accounts of survival in extreme environments, although it does of course happen.[44] Extreme environments may not be the most conducive places for sex: freezing temperatures, high altitude, or microgravity can make it difficult or downright dangerous.[45] And yet people do manage, even in the most unpromising situations.

In 2001, a journalist interviewed veteran climbers about their experiences on Everest. None would name names, but many acknowledged that sexual activity was not unknown on the mountain. One climber knew of a couple that 'made it happen' at the South Col at 26,000 feet, commenting that sex at high altitude was not difficult 'as long as you have supplementary oxygen'.[46]

The dark, dank subterranean environments coveted by extreme cavers do not always dampen their ardour. One author, writing about an expedition to Chevé Cave in Mexico, one of the world's deepest, noted that plastic ground-sheets 'had their drawbacks when it came to romancing on stone. The distinctive crinkle-crinkle sound they made, impossible to ignore, was Chevé's equivalent of creaking bedsprings'.[47]

Historically, explorers were sometimes accompanied, covertly or overtly, by partners. Some eighteenth- and nineteenth-century seafarers smuggled their mistresses on board disguised as male servants. Arctic explorer Robert Peary openly took his wife on one of his expeditions at the turn of the twentieth century.[48]

The limited opportunities for sex, combined with a high ratio of males to females, can create hazards for women on extreme missions. Sexual assaults are not unheard of.[49] According to one mountaineer, young women climbers are prey to 'lecherous, tongue-dragging, testosterone-riddled male climbers'.[50] The world of caving is similarly said to feature 'lots of alpha males jockeying for female attention',[51] and women on Antarctic stations have to develop strategies for dealing with physical and psychological harassment.[52] Men who are successful in extreme environments sometimes attract groupies and succumb to temptation on their return, despite being in committed relationships. As one mountaineer put it, 'having the odd shag here and there seems a lot less dangerous than what you're doing on the mountain'.[53]

What of those who have no opportunity for sex or romance? Those who engage in extreme activities sometimes do so at the cost of enforced celibacy. Apsley Cherry-Garrard wrote that 'both sexually and socially the polar explorer must make up his mind to be starved'.[54]

The medical officer on an Antarctic expedition in 1960 noted that one of the problems of being confined for a year was the 'absence of many accustomed sources of gratification, both sexual and gastronomic'. The doctor added that: 'Interest in sexual topics appeared to be most pronounced at two periods, one in the very early stages, and the other towards the end, when watching the breeding behaviour of the elephant seal became a favourite pastime'.[55] More recently, men who had overwintered at the McMurdo research station in Antarctica joked that they suffered from 'DSB' (deadly sperm build-up).[56]

Most people in situations of enforced celibacy rely on masturbation, although this universal human response is rarely acknowledged and seldom researched.[57] British polar bases reportedly hold libraries of pornographic videos and magazines, maintained by volunteers known as 'Z-porn'.[58] Two veteran Russian cosmonauts spoke with unusual candour when describing how they had spent months together on the Mir space station. The authorities had vetoed the idea of taking along an inflatable sex doll, on the grounds that its use would have to be written into their daily schedule. The cosmonauts found another solution. As one put it: 'We have a joke. You know we have food in tubes. There are white and black tubes. On the white is written BLONDE. On black one, BRUNETTE'.[59]

Some rare statistics were reported in a study of Indian scientists on an Antarctic base in 2000–2001. The great majority reported masturbating, with frequencies ranging from a few times a month to more than once a week, even though 40 per cent of them believed masturbation to be wrong. One in three of the scientists reported that sexual thoughts often interfered with their work.[60]

Psychiatrist Glin Bennet considered sexual behaviour in his study of thirty-four men who were competing in a 1972 solo yacht race across the Atlantic. One yachtsman took along erotic material, but the others apparently felt this was not a good idea. When interviewed by Bennet after the race, none of the competitors admitted to feeling deprived of sex, claiming that the work was all-consuming and any sexual thoughts were short-lived.[61] Even if true for those sailors, other accounts paint a different picture.

For instance, Robin Knox-Johnston's memoir of his solo circumnavigation, based on contemporaneous logs, is peppered with references to dreaming about and yearning for female company.[62]

The strategy used by Shackleton's crew, and by Thor Heyerdahl's raft-mates on their *Kon-Tiki* voyage, was simply to ignore sexual thoughts and pretend that women did not exist—although whether this actually worked is not recorded.[63] Wilfred Thesiger, who spent years walking across deserts with a few male companions, wrote: 'I might have been homosexual if I was born in a different age but as it was I remained asexual'.[64]

In everyday life, some people experience prolonged and involuntary celibacy—they desire sex but lack a sexual partner or are in a sexless relationship. Most find the experience distressing. Researchers found that more than a third of involuntary celibates who had not had sex for six months or longer were dissatisfied, frustrated, or angry about their lack of sex. Some were severely depressed and had suicidal thoughts.[65]

Couples and families

Life in hard places can place strains on interpersonal relationships, but it can also foster lasting friendship and even romance. A cliché of many Holly-wood movies is that danger and other forms of intense experience can stimulate strong interpersonal attraction. The idea is that when individuals are aroused by danger they are more easily aroused in other ways, including romantically and sexually. Disappointingly, there is relatively little scientific evidence to support this view.[66]

A shared passion for extreme activities can certainly bring couples together. For instance, the British explorer Ranulph Fiennes' first wife Ginny was a deep-sea diver and noted explorer in her own right, and the first woman to be awarded the prestigious Polar Medal.[67]

Sometimes, an existing romantic relationship draws one partner into becoming an 'accidental adventurer', as in the case of the Darlingtons. In the immediate aftermath of World War II, US Navy pilot Harry Darlington had signed up for an Antarctic research expedition under the command of Finn Ronne. During the weeks before his departure, he wooed and married the young Jennie. Their honeymoon was the voyage to Chile, from where the expedition would travel to Antarctica. Jennie Darlington did not leave the ship in Chile, however. She and Ronne's wife Jackie became the first two women to overwinter in Antarctica. Just before the end of the expedition, Jennie became pregnant, the first woman to do so on that continent.[68]

Until relatively recently, romantic relationships were largely ignored by organizations planning extreme missions, as though they simply did not happen or were somehow irrelevant. When they did crop up, they were often viewed with suspicion. In the 1990s, for example, two Shuttle astronauts managed to keep their marriage a secret until it was too late to find replacements, prompting NASA to place an outright ban on couples working together on space missions.[69] However, as many experts have pointed out, there are benefits in having well-established couples on long-duration missions. Happy couples can rely on each other for social and psychological support and to avoid sexual frustration.[70] But they are also a potential focus for conflict with others in the team.

In a few cases, whole families have ventured into extreme environments. One example is the Sverdrup 2000 expedition, which consisted of three couples and a 2-year-old child. The expedition, which was studied directly by psychologists, provided a rare opportunity to compare a team's contemporaneous weblog account with a detailed psychological assessment.[71]

The 1999–2000 expedition was intended to mark the centenary of Norwegian explorer Otto Sverdrup's journey from Norway to Canada. It was the brainchild of adventurers Graeme and Lynda Magor, who believed that Sverdrup's achievements should be brought to a wider audience. (Conspiracy theorists believed there was another reason for the trip: the team was made up of couples who were supposedly destined to keep the human race going in case the Millennium marked the end of civilization.) The plan was to retrace Sverdrup's footsteps. They would overwinter, locked in ice in Hourglass Bay off the coast of Ellesmere Island, and then, once the weather improved, travel north by sledge to Bukken Fjord in Inuit territory.

The Magors recruited two other couples to their team. Norwegian biologist Guldborg Søvik and her meteorologist partner Lars Robert Hole embraced the opportunity to study Arctic fauna and weather. The other couple, Greg Landreth and Keri Pashuk, owned the vessel *Northanger* on which they would all live. The team was not selected on the basis of personality tests—or indeed any measure of compatibility—and the couples did not know each other before the expedition. The seventh member of the team was 2-year-old Keziah Magor. If the Magors had any doubts about their child spending months in the darkness of an Arctic winter, those doubts remained hidden.

The lead psychologist who studied the team was Gloria Leon. Her professional interest, on behalf of NASA, was how couples would get on in long-duration missions. With a possible manned mission to Mars on the (distant) horizon, NASA had jumped at the opportunity to study real couples sharing an extreme experience. The team members filled out various questionnaires during the expedition, giving details of what they had been doing, how they were feeling, how they were getting on with their team mates, how much intimacy (of all types) they were experiencing, and so on. In their last week together they were visited by Leon, who interviewed each of them privately about their experiences.

The psychological research and team diaries reveal that expedition members were happier more often than they were stressed and sad. But they had their problems. Being stuck in a small, ice-bound boat during six weeks of Arctic winter darkness was stressful. Although they built a small hut on the shore to extend their living space, the quarters were cramped. Those who spent time on the boat struggled with lack of privacy, finding that conversations carried through the thin internal partitions.

Danger surrounded them. They lived in fear of polar bear attack. The pressure of the ice made them anxious about the boat being damaged—and with good reason, it turned out. A few weeks after Hourglass Bay froze over, they realized that *Northanger* was moving. The boat was gradually tilting and, if unchecked, would flood. Preventing this was a Herculean task involving weeks of hard labour digging through the ice.[72]

And what of little Keziah? The Sverdrup 2000 website featured a question-and-answer page, on which many questions were asked about Keziah, both before and during the journey. The Magors reassured their correspondents that Keziah was well, enjoying the Arctic, playing in minus 35 degrees Celsius, sledging, making snow angels, and eating pancakes. Her third birthday was celebrated with a Paddington Bear-themed party and an excess of chocolate. But although the website gave no sign of it, the psychologists found that the presence of a child was stressful for some team members, who thought her father spent too much time with the girl and not enough on team tasks. One of the other men reported that having a young child on board was one of the most stressful aspects of the expedition, making it difficult for the adults to have ordinary conversations and forcing them to listen to the child's tantrums.[73]

The people left behind

What of the family members who do not accompany their loved ones to hard places? Life is tough in extreme environments, but it can also be tough for those left behind. They must cope with the adventurer's absence, while living with the chronic anxiety that their loved one might be harmed or never return.[74]

Before the arrival of satellite phones and the internet, families could endure weeks or months of worry, wondering whether their loved one was safe. One climber's wife even admitted to feeling a moment of relief when she heard that her husband had been killed, thinking that at least she would no longer have to worry all the time about him dying.[75] The climber Robert Macfarlane argued that mountaineers who keep on taking big risks must be 'either profoundly selfish, or incapable of sympathy for those who love them'.[76]

Being the partner of an extreme adventurer can be emotionally complex. Maria Coffey, the girlfriend of mountaineer Joe Tasker, spoke of how she was 'thrilled by the pace of his life, and relished the vicarious excitement, the reflected glory and the touch of glamour'.[77] Sadly, Tasker died on Everest in

1982. Ruth Seifert, the wife of another Everest climber, found it less thrilling. She was proud of her own career as a psychiatrist and irritated by other people's assumption that her husband had the harder job.[78]

Adventurers who leave behind dependents sometimes rouse the ire of the media. When the climber Alison Hargreaves died on K2 in 1995, several female journalists erupted with moral outrage, attacking Hargreaves' decision to climb such a dangerous mountain when she had two small children at home. One accused her of 'reality-denying self-centredness' and asked: 'If the Alison Hargreaves of this world really value life so little maybe we should not worry on their behalf if they lose it?'[79] And yet for Hargreaves it was no vainglorious expedition. She was a professional mountaineer and the family breadwinner. Her ascent of K2 was her chance to secure her children's future. Neither was she reckless: she was a talented climber who had made numerous first ascents. As a mother, though, she was an easy target for commentators who did not understand the complex reality behind the headlines.[80]

The individuals who travel to extreme environments also find the separation difficult, of course. Many adventurers have written about how hard it was to leave their loved ones behind. Before he set off on his final Everest expedition in 1924, George Mallory wrote of the pain he felt leaving his beloved wife Ruth.[81] The Arctic explorer Adolphus Greely longed to see his wife and children: 'I think of you always and most continually', he wrote. 'There seems so little outside of you and the babes that is of any real and true value to me'.[82]

In 2006, Australian adventurer Andrew McAuley tried to kayak from Australia to New Zealand across the Tasman Sea. The video diary of his first attempt shows McAuley fighting to control his emotions as he paddled away from his wife and child: 'I'm really worried I'm not going to see my wife again and my little boy', he said, his voice breaking. 'And I'm very scared... I have a boy that needs a father and a wife that needs a husband. And I'm wondering why I am doing this. And I don't have an answer'. Towards the end of his month-long second attempt a few weeks later, McAuley was almost at the limits of his endurance. In his video diary, he talked of the pain of salt-water sores, the aching exertion of paddling, and the nerve-wracking storms that battered his cramped kayak. But it was being absent from his family that he found most testing. Despite fierce storms and several near-fatal capsizes, McAuley managed to get to within 30 miles of his destination before disappearing. His body was never found.[83]

In conclusion, social conflict can be a powerful stressor, especially when a small group of people is confined together for long periods in difficult circumstances. The ability to get on with other people is therefore often critical in extreme environments. To survive and thrive in demanding situations, as in everyday life, we must learn to be tolerant and tolerable.

8

Teamwork

They must be able to live together in harmony for a long period without
outside communication...men whose desires lead them to the untrodden
paths of the world have generally marked individuality.

Ernest Shackleton (1874–1922)

Having looked at the conflicts that arise when people are cooped up
together, we turn to another crucial aspect of interpersonal relationships: how people in hard places work together as a team. Teamwork means
more than just getting along with others, although that is an important
element. In a successful team, individuals must work together to achieve
common goals, drawing on a mix of interpersonal and task-related skills. In
this chapter we explore the nature of good teamwork, how to select the
'right stuff' when forming a team, how extreme teams are strengthened, the
role of leaders, and the relationship with ground control.

Being a team

Achieving goals in extreme environments requires strong teamwork, even if
the goal is just to survive. A team that finds itself in dangerous or unfamiliar
situations must respond rapidly and collectively. Good teamwork can be a
matter of life and death, whether for a two-person climbing team, a polar
exploration crew of dozens, or a space mission managed by an organization
of hundreds. Even solo operators need support from their base.[1]

Teamwork is ubiquitous in everyday life, and several decades of research
has explored its nature. Until relatively recently, understanding was largely

derived from research in business or military contexts, but teamwork in other types of demanding environment is now an active area of research.[2]

Fundamentally, teamwork has two broad aspects. The first relates to *what* the team does, and encompasses task-specific skills and knowledge, such as knowing how to drive a motorized sledge in the Antarctic or apply decompression tables when diving. It is sometimes referred to as 'task work'. (We return to task-related skills in Chapter 9, where we look at the vital role of expertise.)

The second aspect of teamwork relates to *how* the team performs its tasks. This encompasses the skills, knowledge, and abilities required to coordinate actions with other members of the team. It includes behaviours and attributes, such as communication and interpersonal sensitivity, which affect how individuals work together in performing a task. No matter how expert the individuals in a team may be, they will not achieve their goals if they cannot work together harmoniously.

At its core, strong team performance is underpinned by coordination, cooperation, and communication. All three are particularly difficult to achieve in stressful and uncertain situations.[3] To understand how teams perform, we must look at what goes on in the minds of team members. And to do that, we need to explain an important theoretical concept: the team mental model.

A mental model can be thought of as a temporary, hypothetical structure constructed in the mind to represent and make sense of external reality. Mental models play an important role in cognitive psychology, as they help to explain how complex decisions are made: we construct mental models to help make sense of information as a prelude to working out what actions to take. Constructing a mental representation allows us to assimilate information from our immediate perception and integrate it with existing knowledge from our long-term memory. We rely on mental models for a number of cognitive operations, such as planning, inferring causes, or predicting future states.[4]

Creating your own mental model of a situation, how that situation might develop, and how you might respond, is hard enough. But what happens when you are one of a team of several individuals, each with their own mental model? The answer is the team mental model: a shared understanding of the information that everyone needs to achieve the team's goals.

Not everyone in a team needs to know everything that is relevant to the task at hand. But there are some types of information they do need to share. For instance, they all need to know the team's goals, the tasks that must be performed to achieve those goals, and the obstacles to overcome. Each team member must also understand what the other people know, what their skills are, how they tend to work, and how the team as a whole works together.[5]

Research suggests that a team of skilled individuals who share a mental model that includes all these elements is more likely to perform effectively.[6] This is particularly true in complex, fast-moving situations where there is no time to stop and ask other team members. Effective teams rely on what has been called implicit coordination, which means operating harmoniously, anticipating each other's actions, and sharing a common understanding of what everyone must do.[7]

Whereas coordination is about acting in synchrony, cooperation is about the way in which those actions are performed. In teams that work cooperatively, individuals monitor each other's performance, noticing when tasks are not distributed evenly or when one person is overstretched. Team members act as each other's backup, stepping in where necessary to share the load. As with coordination, a team mental model that includes knowledge of how other members work, and what their limitations might be, is crucial for cooperation.[8]

Stressful conditions usually prompt a strong team to pull together, but they can have the opposite effect on weak teams. This happened during an extreme caving expedition in Mexico in the 1990s. Psychologists studying the team reported that cooperation among team members declined as they suffered the cumulative effects of stressful environmental conditions, interpersonal conflict, and the deaths of two team members. As the group broke down, some team members mutinied against the leader, while others left the expedition altogether.[9]

The third ingredient of effective teamwork is good communication. In dynamic and uncertain situations, teams must frequently update their shared mental model to accommodate changes. One important advantage that a team has over an individual is its greater capacity to spot changes and new dangers. But unless the team members rapidly share new information, their mental models will start to diverge, the value of new information will be reduced, and the team's decision making will degrade.

One aspect of a team's shared thinking that could be considered part of its broader mental model is the 'team climate'. This refers to the general ethos of the team, which helps shape the way in which it achieves goals. In the business world, for example, a retail team might have a 'sales climate', where success involves pushing their product as hard as possible, or a 'service climate', where ensuring that the customer is happy is more important. A team of extreme operators might have a 'risk-taking climate', where pushing the boundaries is encouraged, or a 'safety climate', where caution is emphasized.[10]

The leader has an important role in setting the team climate, but it is a property of the group as a whole rather than of individuals. Where team members do not share the same team climate, cohesiveness is likely to be undermined unless individuals ignore their own preference in favour of the group. A sole risk-taker in a safety climate team, for example, will tend to adopt the safety ethos or leave the team.

When the team climate is one of competition, the consequences can be damaging. Research on commercial Everest expeditions found that individual climbers tended to focus on their own accomplishments and on achieving unique positions, such as being the first person in a particular category to summit.[11] While competitiveness may not be a problem when the going is good, in crises it can seriously undermine people's chances of surviving.

Shared mental models help us to understand the cognitive underpinnings of teamwork. But we also need to look at the affective (emotional) aspects in order to understand why team members are motivated to work together in the first place. A crucial factor—some would argue *the* crucial factor—is group cohesion.[12] This is a measure of the degree to which members feel bonded to each other, to the group as a whole, and to the work of the group. Group cohesion can be thought of as the glue that binds the group together and motivates them to work with each other.

Research shows that team performance is strongly correlated with team cohesion, and the more complex the task, the stronger the correlation.[13] Although it is not altogether clear whether good cohesion causes good performance or vice versa, the emerging consensus is that the two are mutually reinforcing: high performance makes individuals proud to be in the team (thus greater cohesion) and more inclined to work hard to achieve team goals (thus greater performance).[14] Overcoming difficulties and achieving demanding goals tends to strengthen individuals' belief in the capability of the team, which in turn leads to better performance.[15]

Homogeneous teams, in which the members are similar to each other, tend to be more cohesive than diverse teams. As we saw in the previous chapter, nationality, gender, and professional differences can underlie social conflict in extreme environments. But even when team members are outwardly different, they can still form a cohesive group. Apparently diverse teams can bond over shared experiences, outlook, or values. A common view of life in general, or of the task they are engaged in, can bind members together.[16] In one study, psychologists found that two of the most significant factors behind good crew communication in space were common experiences and a shared excitement about being in space.[17]

Extreme teams tend to be exclusive clubs, where high levels of expertise are necessary to qualify for membership. Where a newcomer's incompetence could prove fatal, existing members think carefully about who should join their team. Individuals may be thrown out, or leave of their own accord, if they do not measure up. Inevitably, such teams tend to end up with people who are similar to each other. This process of homogenization is known as the 'attraction-selection-attrition' mechanism: you attract people who are like you; you select people who are like you; and you lose people who are not like you.[18]

Another factor that tends to strengthen group cohesion is the presence of an identifiable and potentially threatening 'out-group'—that is, people who are not part of the group and do not share its group identity.[19] Individuals who engage in extreme activities such as mountaineering or cave diving sometimes face stinging criticism from those who question their behaviour on ethical, financial, or practical grounds, often with varying degrees of ignorance. Such criticism can strengthen the commitment of those within the community of practitioners.

Cohesion can have its downsides, however, such as when it contributes to 'groupthink'. This occurs when a highly cohesive team under pressure develops a form of collective tunnel vision and fails to see flaws in its plan.[20] A related pitfall is group polarization, whereby the team as a whole develops more extreme attitudes (including, potentially, towards risk-taking) than would be predicted from the attitudes of individuals.[21]

Cohesive teams can also fall victim to other biases such as false consensus, where individuals overestimate the degree to which others in the team share an understanding. This can be dangerous in extreme situations where conditions are changing rapidly and communication is difficult. Individuals may make false assumptions about what their teammates are thinking or doing, and fail to examine those assumptions because they feel right.[22]

Consequently, some degree of tension within a team is not always a bad thing. Challenging others about their plans and potential courses of action can be helpful in uncovering new information or risks that might otherwise be overlooked. On the other hand, social conflict, of the kind we discussed in the previous chapter, can be highly damaging. Social conflict tends to lower the overall mood of a team, even among members who are not directly involved in the conflict, and depressed mood often impairs performance. Team members may then find themselves having to deal with rebuilding relationships rather than getting the tasks done. If a feud simmers, this can reduce cohesiveness and impair the team's ability to succeed.

Organizations that run extreme missions try to monitor team cohesion, giving them early warning that trouble may be brewing. One way of doing this is by monitoring language use. Cohesive teams tend to use collective words like 'we' and 'us' rather than 'I' and 'you'.[23] According to one study, phrasing instructions in collective terms—such as, 'we should do this' rather than 'do this'—is more likely to result in team members correcting errors.[24]

Before we leave the subject of team cohesion, we will look at an issue that has long troubled the organizers of extreme missions. That issue is gender.

Happily, things have moved on since 1982, when female cosmonaut Svetlana Savitskaya visited the Soviet Salyut 7 space station for an eight-day mission. She was handed a flowery apron and a bunch of flowers and asked: 'Look, Sveta, even though you are a pilot and a cosmonaut, you are still a woman first. Would you please do us the honour of being our hostess tonight?'[25]

Mixed gender teams are now common in all extreme environments and the research shows that they are, mostly, a success. A growing body of evidence suggests that having a mix of males and females on long-duration missions helps to normalize behaviour.[26] All-male groups tend to be competitive, but the presence of women (who tend to take the role of peacemakers) can reduce the competition and associated tension. Some researchers argue that women are more inclined to build and maintain social bonds as a buffer against stress.[27]

Mixed teams might be good for men, but women can find them more problematic. Women can become the targets of sexism, as in the case of cosmonaut Savitskaya, or of well meaning but unhelpful chivalry. Studies of small groups in real and simulated extreme situations have found that even when overt sexism is not apparent, men in mixed teams often treat women differently—for instance, by confiding their concerns to a woman, although without being good at listening in return.[28] Might women be better off in all-female teams? Some studies have suggested that cooperation is greater in all-female teams compared to mixed or all-male teams. But that is not always the case.

When Arlene Blum led a twelve-woman team in the ascent of Annapurna in 1978, what started out as a friendly, democratic expedition disintegrated on the mountain. The team had, unusually, worked with a psychologist prior to their trip and agreed that decisions should be taken democratically. But their good intentions withered as the expedition proceeded.

When Blum tried to be decisive, the others complained about not being included. When she sought everyone's input, they disagreed among themselves and made her look weak as a leader. Squabbling about the division of labour between each other and the (male) Sherpas further soured the mood. Nerves were on edge: terrifying avalanches narrowly missed killing team members and obliterated caches of food and equipment. Blum was reminded that more people had died trying to reach Annapurna's summit than had actually got there. She no longer had the ambition to reach the top, but other team members remained eager to try. It was impossible for all the women to make an attempt on the summit, and Blum's choice of who should go created more disagreement. Two of the women she chose did reach the summit, along with two male Sherpas. However, two women she had tried to dissuade died in their attempt.[29]

Other teams of women have fared better. Psychologist Gloria Leon and her colleagues studied two all-female teams that had undertaken gruelling polar expeditions during the 1990s. The first was a two-woman team who trekked across the Antarctic. The second was a multinational four-woman team who took six weeks to traverse Greenland.[30]

The two women in the first team, Ann Bancroft (an American) and Liv Arnesen (a Norwegian), were very similar in measures of personality. In particular, they were both low in 'verbal aggressiveness' and high in 'positive expressivity', as measured by the Personality Characteristics Inventory. They had similar motivation: a love of the Antarctic, experience of crossing the continent, and a desire to be good role models for women and girls. The two women made decisions and solved problems in similar ways. They coped with hardships using a variety of positive strategies, including humour, and got on exceptionally well—so well, in fact, that Arnesen claimed to have found, in Bancroft, her 'sister soul'.[31]

Despite only meeting as a team for the first time a fortnight before the expedition, the four-woman multinational team also functioned well. Apart from bad weather, most of the stresses reported by the team were social in nature, but these were focused on the well-being of others and the need to maintain positive relationships. Language difficulties were frustrating: three of the four (a Greenlander, a Dane, and a Brit) spoke good English, but the Russian struggled. They all got homesick, cold, and tired, and by the halfway point were struggling with their motivation. But they worked hard to maintain positive relationships and used body language where necessary to communicate with the Russian woman. In return, she taught them Russian songs.[32]

Psychologists found that the women in the Annapurna expedition displayed characteristics that were reportedly similar to exceptional male climbers—in particular, self-sufficiency and reserve.[33] Such traits may help individuals reach a summit but are not necessarily conducive to teamwork. The six women in the polar exploration teams, on the other hand, shared personality traits that were valuable for being a team. These examples suggest that the ability of a team to get along may be as much about individual personalities as the gender or other socio-demographic characteristics of team members.[34]

All but one of the six women in the two polar teams had personality profiles that psychologists described as the 'right stuff'—the set of personality characteristics that research shows are most likely to be valuable in hard places.[35] What exactly are these characteristics?

Choosing the right stuff

In the late 1950s NASA psychiatrists were given the novel task of assessing the first US astronauts for the Mercury space programme. The received wisdom had been that the men who volunteered for these dangerous missions would be 'impulsive, suicidal, sexually aberrant thrill-seekers'.[36] This turned out to be far from the truth.

The NASA psychiatrists subjected the candidates to batteries of psychological tests and interviews. Some astronauts engaged with interest; others viewed the tests as an unnecessary annoyance. The tests produced a wealth of data, but they were classified ('for reasons that we never understood', remarked one of the psychiatrists later).[37] Before the publication of their research was halted, the psychiatrists released a couple of papers in which they detailed the personalities, developmental histories, and pre- and post-flight reactions of the first astronauts.[38]

These papers portray the men who made up the Mercury crews as near perfect specimens of humankind: highly intelligent, practical, action-oriented, and aware of their own emotions but in control of them. These men were not seeking danger for its own sake, but they were prepared to accept it as part of their mission. They appeared strongly motivated by the need to master the situations they were in, and their thorough training, combined with a long history of achievement in dangerous situations, gave them supreme confidence that they could overcome the challenges of spaceflight. When things went wrong, they were resilient in the face of frustration.[39]

To a man, the astronauts reported having stable families, a strong father figure, and a smooth, steady progression through the various developmental stages to adulthood and later in their chosen professions. The psychiatrists speculated that the astronauts' upbringings had given them the firm identities and solid skills needed to cope with extremely testing situations.[40]

The men appeared perfect. But could they really have been such paragons of virtue? Former NASA psychiatrist Pat Santy, who wrote a classic account of astronaut selection, was sceptical.[41] She wrote: 'This is a group of individuals who are hardly motivated to voice psychological problems because of the consequences of doing so'. Santy commented that the astronauts regarded their interactions with medical professionals as 'battles they could never win'.[42] As Tom Wolfe later put it in *The Right Stuff*, the astronauts were frustrated at having 'too many goddam doctors involved in this thing'.[43]

The NASA psychiatrists had encountered a problem that is common in selection. When selecting individuals from a pool of candidates who are all highly motivated to be chosen, how do you know that they are truly skilled, resilient, and compatible? More importantly, how do you ensure that highly intelligent candidates will give candid answers when they know that the wrong answer will end their dreams? The solution is not to rely on self-reporting, but to put candidates through their paces in simulations and other objective assessment exercises.

Those involved in recruiting personnel for extreme environments talk of 'selecting in' and 'selecting out' processes.[44] In the former, applicants are assessed against a standardized set of personality characteristics and behavioural competencies that are believed to make them suitable for tough situations. This set of qualities is sometimes referred to as 'right stuff'. In the 'selecting out' process, individuals are removed from the pool of candidates because they possess 'wrong stuff'—attributes that would be harmful or likely to result in poor performance.

Nowadays the selection process for astronauts can take several years, as space agencies whittle down thousands of credible applicants to a handful who have the requisite mix of technical expertise, but also—crucially—a stable and agreeable personality, and interpersonal competence. As astronaut Chris Hadfield put it: 'No one wants to go into space with a jerk'.[45]

Candidates are assessed through lengthy interviews with psychiatrists or clinical psychologists, who look for evidence of instability or early signs of potential mental health issues. This form of screening is considered essential by most organizations that select personnel for extreme environments, including space agencies and polar research institutes. But because they are selecting from such a highly capable pool of candidates, the proportion that get screened out tends to be rather small. The selectors are then left with a large pool from which to 'select in'.[46]

What are the qualities that should be 'selected in'? When choosing crew for his expeditions, Ernest Shackleton noted that 'their science or seamanship weighs little against the kind of chaps they were'. For the aptly named *Endurance* expedition of 1914–1916, Shackleton started with a core of preselected men with whom he had previously worked and upon whom he could confidently rely. The thousands of applications for the remaining places were sifted into 'mad', 'hopeless', or 'possible'. Shackleton then personally interviewed those in the third category. He looked for men who were cheerful, shared his vision, had the skills he needed, and were unafraid of hard work.[47]

The *Endurance* expedition ended with one of the most famous feats of survival in the history of exploration. Shackleton and his crew were ice-locked for almost a year until their ship was crushed and sank. They spent four months of Antarctic winter in flimsy tents before setting out in three small lifeboats in search of rescue. Shackleton left most of his crew on a barren island, where they endured near-starvation, while he and five others made a desperate 800-mile journey in an open boat to South Georgia, culminating in a three-day trek across the mountains to reach help. All survived. What is more, they remained remarkably cohesive and in reasonable spirits throughout their ordeal.[48]

Shackleton clearly knew what he was doing when selecting his team. A century later, many recruiters for extreme expeditions use more sophisticated methods grounded in the scientific study of personality characteristics. But their underlying approach is essentially the same, because research has broadly confirmed the validity of Shackleton's intuitive method of selecting good team players.

The research indicates that the relationship between individual personality traits and team effectiveness is not straightforward, and depends on the tasks the team is required to perform. Moreover, it is not just the qualities of each individual that matter, but also the overall mix. Agreeable and sociable extraverts may be part of the social glue that binds the team together, but the conscientious and detail-focused planners are needed to ensure that the tasks are completed.[49]

Some broad conclusions can be drawn, however. First, team members need to be emotionally stable, particularly for extreme environments. Individuals who are low in measures of emotional stability tend to be moody, anxious, and neurotic worriers who brood rather than take positive action in the face of setbacks. Insecurity makes them more inclined to blame others for their mistakes rather than taking personal responsibility. They may not pull their weight and can quickly start to get on others' nerves. Furthermore, bad moods can be infectious: one neurotic individual can bring down a whole team.[50]

Emotional stability and self-confidence are both desirable qualities, as long as they are not accompanied by a preference for dominance. Domineering individuals, especially if there is more than one, can destabilize a team by fostering conflict and pushing others into submissive roles. Too much compliance could leave the team so reliant on the dominant individual that it cannot function if that person is lost, for instance, in an accident.[51] Where there are two dominant individuals, friction is almost inevitable. A team should not have two leaders.

A third important quality is flexibility. Team members who can adapt readily to changing circumstances will perform better than those who are rigid, stubborn, and unwilling to change.[52]

A further set of desirable qualities is related to how team members think about and behave towards each other. We touched earlier on the importance of having a cooperative (as opposed to competitive) orientation. Team members must also trust each other, as well as being trustworthy and reliable themselves. In extreme environments, you may have to trust your teammates with your life.

Several established personality tests include ways of measuring an individual's tendency to trust or be suspicious of other people. Those who score highly on measures of trust tend to believe that others can be relied on unless there is evidence to the contrary. People who assume that others are basically trustworthy also tend to be more cooperative, and, as we have seen, cooperativeness is a good thing in teams.[53]

The mechanisms by which trust increases team performance are not entirely clear. There could be a direct relationship: more trusting team members might be more cooperative, for instance. The mechanism could also be indirect, with high levels of interpersonal trust enabling team members to stop worrying about interpersonal issues and focus instead on the task at hand.[54]

Pockets of mistrust can undermine the performance of an entire team. In one study, researchers measured levels of trust and performance in more than fifty real-life healthcare teams. They found that teams with higher levels of overall trust performed better than those with lower levels, but only when all members of the team trusted each other. When members trusted some individuals more than others, the overall level of trust made no difference to performance.[55] In other words, mutual and equal trust between team members is what matters, not just a broad overall level of trust that could mask pockets of mistrust.

Trusting others is sometimes assumed to be a generalized trait. But it is possible for someone who is generally suspicious or cynical about other people to have high levels of trust in particular individuals.

A final element in the set of interpersonal competencies is the extent to which individuals are 'agreeable'. Psychological measures of agreeableness tap into a person's tendency to feel warm towards, and get on with, other people. Trust is part of this characteristic, as are kindness, empathy, and tolerance towards others. The opposite qualities, including intolerance, coldness, and selfishness, are obviously not those of a team player. Several studies have found that measures of agreeableness are predictive of coping in extreme environments. For instance, people who are friendly and participatory fit in better and feel happier in polar stations, according to a study of more than a hundred people who spent time in Antarctica.[56]

Team training

Having selected a skilled and compatible team, the next step is to prepare them for their extreme experience. Team members with specific roles may need to work on their own specialist skills, but the team as a whole should also train jointly.

When a team trains together it may focus on the specific skills needed for task work, or on improving teamwork. Research on the effectiveness of different forms of team training shows that almost any sort of training will improve team performance, largely by helping the team to achieve a clear and shared understanding of their goals and what they must do to achieve those goals. Team-building exercises, although often mocked, can in fact help to promote cohesion and thereby improve teamwork.[57] However, the greatest benefit accrues when the training (whether task- or team-focused) matches the needs of the team. A cohesive team preparing for an extreme environment might enjoy a team-building exercise, but it will not do them as much good as task-related training.[58]

Where teams are multinational, they can further benefit from pre-mission training designed to enhance their cross-cultural awareness and build familiarity between the soon-to-be crewmates. In the 1990s, US astronauts learned Russian and lived in Russia for a time to familiarize themselves with the culture and customs of the cosmonauts with whom they would share the Shuttle-Mir missions. The failure of the Russians to prepare in the same way (largely because of costs) is believed to have led to problems during those missions.[59]

Simulation exercises, in which tasks are performed in a realistic training environment, can be effective in preparing a team for its mission and identifying areas for further development. A crucial feature of any simulation is its psychological fidelity—that is, the degree to which the simulation tests the same emotional, cognitive, and social skills as the real situation.[60] A simulation can test a team's ability to handle a crisis and give them a taste of living together in close proximity for long periods.[61]

Psychological fidelity is further improved if the simulation replicates some of the danger of the real environment. To that end, the European Space Agency sends multinational crews deep underground to gain experience of working as a team in a genuinely hostile environment. The 'cavenauts', as the Agency calls them, go so deep that, in the event of an emergency, it would take them longer to get back to the surface than it would take astronauts on the International Space Station to get back to the safety of Earth.[62]

In sum, a successful mission requires a skilled team, selected for compatibility and trained to work well together. But one further and crucial ingredient is also needed. That ingredient is leadership.

Leadership

Leadership is easier to describe than to define. As long ago as 1974, the author of an academic book on the subject commented that there are 'almost as many definitions of leadership as there are persons who have attempted to define the concept'.[63] The field has become much more crowded with experts and definitions since then.

In contrast with the much-lauded Ernest Shackleton, his fellow Antarctic explorer Captain Robert Scott has come in for criticism for his failings as a leader. Apsley Cherry-Garrard's description of Scott reveals something of the man's complexity, referring to him as shy, reserved, 'femininely sensitive, to a degree which might be considered a fault', and subject to dark moods that could last for weeks. Furthermore, says Cherry-Garrard, Scott had little sense of humour and was a bad judge of men. This was hardly glowing praise from a man who had been one of Scott's closest colleagues during his final expedition. Perhaps wishing to balance the criticism, Cherry-Garrard concluded his portrayal by conceding that Scott's greatest triumph was 'that by which he conquered his weaker self, and became the strong leader whom we went to follow and came to love'.[64]

Being a leader in a hard place is not easy. For a start, the people who must be led are unusual individuals with high levels of drive and accomplishment in their own right. The team and its leader must contend with extraordinary stressors that affect their cognition, emotions, and behaviour in ways that are not always obvious. Furthermore, leading in an extreme environment can be a lonely business: there are few people with whom the leader can share their concerns, and no peer or superior to ask for advice or support.[65] As the polar explorer Admiral Byrd noted, the leader is always on show: 'everything one does or says or even thinks, is of importance to one's fellows. They are measuring you constantly, some openly, others secretly—there is so little else to do!'[66]

The leader of an extreme mission is denied many of the tools that everyday leaders use to discipline and reward followers. Withholding privileges or expelling someone is not easy or ethical in an extreme environment. A disobedient employee cannot just be sacked or sent home. Leaders can give rewards, but the opportunities are limited and the rewards, such as breaking out a bottle, must generally be directed at the team rather than one person. Any suspicion of favouritism can be divisive.

A team in an extreme environment needs a leader who is focused on the mission but also alert to the psychological needs of the team. The leader should be able to articulate a vision and motivate the team but not be obsessed with a distant goal. They must be decisive and authoritative, yet sensitive to their followers' limitations.[67]

Military psychologist Thomas Kolditz has researched the qualities required of leaders in extreme situations. Drawing on observations in military and skydiving contexts, Kolditz argues that what he calls 'in extremis' situations require a distinct style of leadership: the leader's responsibility for people's lives means that leadership in hard places is 'less about power over subordinates and more about an obligation toward their well-being and survival'.[68] This has implications for the type of person that makes a good 'in extremis' leader.

The task of an 'in extremis' leader is different to that of a leader in a conventional organization or business. Corporate employees need to be inspired and motivated by their leader to commit to the organization's goals. In contrast, those who choose extreme environments are already highly motivated to be there. The 'in extremis' leader's role is not to persuade the team that their mission is worthwhile, but rather to enable them to strengthen their expertise and teamwork.

To win loyalty and commitment from their followers, leaders in extreme situations cannot rely on badges of authority or hierarchical position. Instead, they must gain trust by establishing and maintaining personal credibility.

What exactly is credibility? Psychologists have found it to be a slippery concept. Some have argued that credibility is not an intrinsic quality that someone either has or does not have, but rather a characteristic that exists only in other people's perception. Credibility, in other words, requires an audience, just as leadership cannot exist independently of followers.

Credibility has no accepted standard definition. However, two broad themes recur in the literature, particularly in relation to extremes. When people judge a leader's credibility, their judgement is influenced by two main factors: competence and trustworthiness.[69] There is no place in an extreme environment for a leader who is incompetent or untrustworthy, however agreeable or charismatic they may be.[70] (We will return to the subject of competence in Chapter 9.)

Studies have shown that trust in the leadership has a strong influence on the behaviour and feelings of team members. It affects their overall performance and commitment, and how they behave towards others—for example, the degree to which they are altruistic and courteous. Team members who trust their leader are much less likely to quit.[71] (Physically quitting in an extreme environment might not be a practical option, but psychologically quitting is always an unwelcome possibility.)

A leader's trustworthiness has two main dimensions: character and behaviour. Good leaders are perceived as reliable and consistent in character, without being rigid. They are fair and honest, do not cynically manipulate others, and do not indulge in playing power games. They are also judged in terms of their actions. An 'in extremis' leader does not ask their followers to do something they would not do themselves, demonstrates concern for the welfare of their followers, and is prepared to put their followers' needs above their own. They feel responsible for and protective of their crew.[72] Ernest Shackleton was an exemplar of these virtues.

Leadership also has a dark side that is sometimes overlooked. The theory of leadership is often taught in terms of the biographies of heroic and successful leaders, while the bad ones are ignored. Yet there are many examples of leaders in extreme situations who were roundly disliked and, in the worst cases, prompted mutiny.

Bad leadership is sometimes described as 'toxic' or 'destructive'. These terms are applied to damaging leadership behaviour that is systematic and repeated, rather than isolated instances of poor judgement. Even the best leaders have the occasional bad day. Where good leaders use influence and persuasion, toxic leaders rely on dominance, coercion, and manipulation. They behave in ways that are destructive for their followers and the organization for which they work.[73]

The US Army has conducted extensive research into toxic leadership. One study, which surveyed a large sample of army personnel in 2009–2010, found that most of them had encountered toxic leaders. The typical characteristics of toxic leaders included pettiness, micromanagement, rigidity, and indifference to the well-being of their troops. Their subordinates often referred to them as 'assholes' or 'jerks'. Despite being loathed by their juniors, toxic leaders often managed to look good to their superiors.[74] In the civilian world, various workplace surveys have found that around two-thirds of employees identify their immediate boss as the most stressful feature of their job.[75]

Toxic leadership needs more than just a toxic leader before it can take root: it also requires susceptible followers and a conducive environment.[76] One academic has argued that toxic leaders attract followers because they satisfy their followers' basic psychological needs for security and predictability in an uncertain world.[77] An environment conducive to toxic leadership is more likely to develop when there is a perceived external threat, instability, and a lack of checks and balances.[78]

Ground control

In extreme environments, the concept of the team often extends beyond the members of the core mission team itself to include the wider network of people who provide logistical and social support from a distance. These people are generally referred to as ground control (or mission control).

The role of ground control may be highly directive, as in the case of space missions where the crew has a degree of autonomy but must follow the instructions of controllers back on Earth. Other, less directive, types of ground control provide logistical support, such as monitoring the weather for a mountaineering team. At its least directive, ground control may consist of a few individuals (perhaps family members) on the end of a communication link, who provide emotional support to the team and update the outside world on its progress.

For a small team in an isolated and extreme environment, ground control can be a crucial means of communication with the outside world. One type of information that is often filtered through ground control is news about family and friends. Such information is valued highly.[79] One large-scale survey of Antarctic winter-over personnel found that, at times, contact from family and friends was valued more than support from fellow crew members.[80]

News from home can be a mixed blessing, however. Messages from loved ones may remind crewmembers of what they are missing, and bad news can dampen morale. News of world events may also have an impact. Cosmonauts who had left the ground as Soviet citizens found it unsettling to return to Earth in 1992 as Russians.[81] On being told of the 9/11 terrorist attacks on the US in 2001, an American astronaut in the International Space Station reportedly 'zipped around the station' until he found a window that would give him a view of New York City.[82]

As well as relaying news, ground control has a direct working relationship with the mission team. Although this relationship is essential, it can be fraught. Sometimes a team's hostility towards ground control seems justified. On one space mission, for example, ground control failed to take account of the crew's need for rest, overloading them with a punishing schedule of scientific and maintenance tasks. Eventually, the crew rebelled and refused to do any more until the schedule was renegotiated.[83]

In other cases, ground control personnel have become scapegoats, as mission teams displace irritation they feel towards fellow crewmembers on to individuals in ground control.[84] In one study, astronauts on the International Space Station recorded measures of their own mood and their perceptions of the support they were receiving from ground control. Critical incident logs were analysed for evidence of actual changes in support. When the astronauts were feeling moody, they tended to judge the support from ground control to be poorer, even though the logs showed there was objectively no less support.[85] Similar findings emerged from studies of the Shuttle-Mir space missions in the 1990s.[86]

The existence of a common enemy can enhance the cohesion of a team. This means that some degree of friction between ground control and the mission team could have benefits. However, deliberately encouraging such friction is not recommended, because it can too easily spiral into a damaging breakdown in the relationship. A team in an isolated and extreme environment cannot afford to fall out with the people on whom they depend for their success and even survival. A good relationship between the mission team and ground control is also better for the emotional well-being of all involved.

The idea that deliberately creating tensions with ground control might be helpful was tested—and found wanting—in the 2010 Chilean mine accident. After many days of desperate searching, the rescue teams made contact with the thirty-three miners trapped deep underground, but it was clear it would take weeks to drill down and save them. A Chilean psychologist therefore devised a strategy to keep the miners sane until they could be rescued. This strategy was based on the assumption that if members of the ground control team became the targets of the miners' frustration, then that might prevent the miners from turning their anger on each other. Good behaviour would be rewarded and bad behaviour punished by giving or withholding treats. The miners were resentful at being treated like children and resisted. One of the ground control team described it as 'a daily arm wrestle'. The strategy had the intended effect of turning ground control into the 'out-group', but it did little for the well-being of the miners or their rescuers.[87]

In most cases, the relationship between a mission team and ground control is relatively harmonious. Crises can sometimes show this relationship at its best. A famous example is the rescue of the Apollo 13 astronauts in 1970. They had planned to land on the Moon, but an oxygen tank exploded two days into the mission, triggering a cascade of debilitating failures including loss of power. The functioning of hundreds of systems, controlling everything from life support to communications and navigation, was dangerously compromised. Over the next four days, the three astronauts and the huge ground control team worked intensively together to ensure the spacecraft's safe return to Earth.

Before the launch, the Apollo crew had practised for months in simulators, handling all manner of extreme situations. Between them, the crew and ground control knew the spacecraft intimately, enabling them to improvise creatively to tackle problems for which they had not practised. Together, they stopped the venting of oxygen into space, worked out which of the hundreds of electrical systems they could shut down to minimize power use without killing the crew, built a makeshift carbon dioxide removal system to halt the gradual asphyxiation of the astronauts, and steered the crippled spacecraft back to Earth. Against all the odds, the three astronauts were saved in an extraordinary display of teamwork and professional skill.[88]

The experience of Apollo 13 and other extreme missions, together with a body of research evidence, suggests that a ground control crew should be considered an integral part of the overall team. As such, they should be subject to the same principles of selection and training that apply to the mission team itself.[89]

As we have seen, a team must have the right mix of personalities as well as the right skills if it is to be successful. Knowing how to perform tasks is not enough: team members must also be able to get on with each other. Building a successful team starts with choosing people who have the 'right stuff' and rejecting those with the 'wrong stuff'. Finally, good leaders must be competent, trustworthy, and genuinely concerned with the well-being of the individuals in their team.

9

Know-how

Adventure is a sign of incompetence.
Vilhjalmur Stefansson, *My Life With the Eskimo* (1913).

Surviving and thriving in hard places requires more than just courage and a cool head. Specialist knowledge and skills are needed, along with thorough planning and preparation for the risks that lie ahead. While it is not unknown for people to venture unprepared into extreme environments, many of them die as a consequence of their recklessness, and those who do survive often owe their lives to luck or brave rescuers. Experts do not gamble their lives on luck: they make sure they are equipped with the right skills and know what to do if things go wrong.

In this chapter we discuss the factors that interfere with good judgement. We explore how people plan and prepare for risky situations, how expertise contributes to good decision making, and how poor judgement can result in catastrophe.

Making good decisions

In extreme environments most decisions are inherently risky. High-stakes judgements are often made under time pressure and in dynamic situations that are difficult to read. There are social pressures too, including the potential loss of personal credibility if mistakes are made. On top of that, the decision maker can fall victim to a range of common cognitive biases and be subject to physical stressors that affect thinking in unhelpful ways.

Even seemingly straightforward decisions can be risky. Their very mundanity can be a hazard in itself. Routine actions need less thought, but in situations where the safety margin is narrow even mundane tasks must be carried out carefully. That is why people who operate in extreme environments check and recheck their kit before they undertake dangerous activities. Even small mistakes can have catastrophic consequences.

The cave diver David Shaw set out in 2005 to recover the body of a diver who had died at the bottom of a freshwater cave in South Africa ten years earlier. At over 260 metres, Shaw's dive was one of the deepest ever made from the surface using rebreather apparatus, which removes carbon dioxide from exhaled breath. Shaw carried a small torch as he entered the cave. He would normally have draped the torch around his neck, but on this occasion he was filming the rescue attempt using a helmet-borne video camera. The camera prevented him from stowing his torch in the usual way.

The video recording was later recovered, making Shaw's death one of the best documented of all diving fatalities. The torch became tangled in a line, and as he struggled to free himself, the physical exertion and stress led to a rapid increase in his breathing rate. However, he could not breathe deeply enough to reoxygenate his lungs and clear them of carbon dioxide. He rapidly lost consciousness and died within a few minutes, probably from acute respiratory failure and carbon dioxide poisoning.[1] A small error—failing to ensure a piece of equipment was stowed away—had fatal results.

What are the features of extreme environments that complicate decision making? First, there is uncertainty. Information is rarely complete or clear-cut, and perceived events or situations may have multiple potential causes or outcomes. Perceptions may sometimes be wrong, and when there is a degree of sensory deprivation the decision maker may fall victim to illusions or hallucinations. In conditions of uncertainty, a decision maker must piece together a picture of what is going on as best they can—a process known as situation assessment. They must also deal with the risk of their assessment being wrong. We shall return to how they do this, and why expert knowledge is crucial to the process, later in the chapter.

Extreme environments are also uncertain because they are dynamic. Circumstances can change quickly and without warning. A caving support team keeps a close eye on the weather above ground, knowing that even a relatively light shower can cause a rapid and dangerous rise in underground water levels. Mountaineers similarly rely on communications with ground control to keep them informed of weather conditions. Fatal storms can blow into Everest at short notice when climbers are deep into the 'death zone' and several hours away from safety.

An accident can change a situation from one of reassuring progress to life-threatening crisis almost instantly. People's responses to an accident have further consequences that can rapidly become part of the problem. As astronaut Chris Hadfield put it: 'No matter how bad a situation is, you can always make it worse'.[2] This happened to Joe Simpson after he broke his leg while climbing an Andean mountain. The injured Simpson was descending, roped to his climbing partner Simon Yates. While lowering Simpson off a cliff, Yates realized that the rope was too short for Simpson to reach solid ground. Worse, Yates was being dragged towards the edge of the cliff. He decided to cut the rope, letting Simpson fall but saving himself. Simpson was left for dead in a crevasse beneath the cliff, the problem of his broken leg now compounded by being stuck, alone, in an almost inescapable location. Remarkably, he survived.[3]

Time pressure is another factor in many extreme situations. Decisions may have to be made very quickly because, for example, the weather is changing, the air supply is almost exhausted, or a crucial piece of equipment has failed. Time pressure may sometimes be illusory. When the stakes are high, individuals sometimes behave as if they are under time pressure even when they actually have plenty of time to make a more considered decision.[4]

The combination of time pressure and high stakes can interfere with decision making, particularly when combined with other stressors. The decision maker may develop tunnel vision, focusing on details and losing sight of the bigger picture. This 'cognitive narrowing' can be helpful when time is tight and every cognitive resource must be focused on the essentials.

However, it can lead to poor decisions when crucial aspects of the situation are ignored.[5]

Making rapid, ill-considered decisions because of time pressure or high stakes can lead to actions that start to look like panic, which is the last thing you need in an extreme situation. Panic results from intense fear, and fear (as you may recall from Chapter 2) has cognitive, behavioural, and physiological components. A person who believes that time is running out (cognitive response) and is struggling unsuccessfully to escape (behavioural response) experiences a further cognitive response ('I can't get out and I'm going to die'). This triggers physiological responses, such as rapid breathing, which can, as in the case of David Shaw, make matters worse. Panic is the most common cause of death among divers.[6]

Panic can also be fatal for cavers, who sometimes get stuck in narrow spaces and die miserably. The French caver Norbert Casteret, whose exploits inspired generations, noted that 'such accidents start with the awful fear of not being able to get out. This causes one to stiffen up and to make superhuman efforts which soon bring exhaustion'.[7] Suppressing panic and relaxing is the best—sometimes the only—way to escape a desperate situation.

Environmental stressors can interfere with cognition and hence decision making. Extreme cold, for instance, affects the capacity to think: reaction times slow down and memory is impaired. The more complicated the task, and the colder the environment, the worse the impairment.[8]

Hypoxia (lack of oxygen) is a pervasive stressor at high altitudes. At the top of Everest, each breath contains only a third as much oxygen as it does at sea level. A lack of adequate oxygen interferes with basic cognitive functions including alertness, motor coordination, and judgement. Someone who is hypoxic may be incapable of carrying out even simple tasks or remembering new information.[9] Hypoxia also interferes with sleep, which (as we saw in Chapter 4) impairs thinking.[10]

Hypoxia killed two French balloonists in 1875, during an ascent that reached 28,000 feet. They had taken with them a small quantity of bottled oxygen, believing it would allow them to survive in the thin air. But all it did was make them euphoric and dangerously confident about ascending even higher until the oxygen ran out. A third balloonist survived, though with permanent hearing loss. He later described the insidious effects of hypoxia in these terms: 'the body and mind weaken little by little...one becomes indifferent; one no longer thinks of the perilous situation or the danger'.[11]

Divers breathing compressed air may also suffer cognitive impairments. Nitrogen narcosis (or 'rapture of the deep') occurs when nitrogen is breathed under pressure. It affects thinking, impairing the diver's ability to perform simple reasoning or motor tasks, and often leaves them disorientated and confused. The diver may experience mild to moderate euphoria (hence 'rapture'). As with hypoxia, the creeping effects of nitrogen narcosis are hard to notice, making deep diving on air an unpredictable and risky activity.[12]

When making decisions about complex tasks, astronauts have to take account of the effects of microgravity (weightlessness) on their fine motor movements. In one long-term study, a cosmonaut's hand-eye coordination was tested repeatedly during a 438-day mission on the Mir space station. It took three weeks of spaceflight for his performance to recover to pre-flight levels. Post-flight tests showed a similar disturbance in performance as he readjusted to normal gravity.[13] Another unpleasant and distracting effect of microgravity is space motion sickness, experienced by up to two-thirds of astronauts. Motion sickness is miserable at the best of times. Fortunately, most astronauts quickly adapt.[14]

Tiredness, whether caused by sleep deprivation, physical exertion, poor nutrition, or sickness is an important stressor that interferes with judgement. It can significantly impair an individual's assimilation of information, as well as their ability to make sense of information, their memory, and reaction times. When judgements are about risk, tiredness can be particularly dangerous.

The evidence shows that as people become more tired, their ability to assess risks and update their judgements in the light of new information becomes progressively impaired. In one study, for instance, French scientists monitored the risk-taking propensities of military pilots while they took part in a maritime counter-terrorism exercise involving 24 hours of difficult flying. As the pilots grew more tired they became more prone to taking risks and more impulsive.[15]

Research has found that tired people become more inclined to take risks that might produce big gains, and less inclined to take risks that might lead to loss or that require effort.[16] Sleep deprivation also impairs their ability to inhibit or withhold inappropriate responses.[17] As if that were not enough, sleep-deprived people tend to perceive themselves as *less* inclined to take risks.[18] In extreme environments, this combination of risk-taking, impulsiveness, and lack of self-awareness is obviously a dangerous cocktail.

Our ability to make good decisions about risk is further affected by our emotions. The pleasurable sensation associated with the anticipation of gain can bias us towards risk-seeking behaviour, whereas the anxiety associated with the anticipation of loss can bias us towards an irrational degree of risk aversion. This distinction is reflected at the neurobiological level. Brain imaging studies have found that risk-seeking behaviour is preceded by activity in a region of the brain associated with the anticipation of pleasurable experiences such as sex, drug taking, and monetary gain. Risk-aversive behaviour, on the other hand, is preceded by activation of a brain region associated with the anticipation of physical pain or unpleasant sensations.[19]

Social pressure is another factor that can interfere with judgement and decision making. Knowing that other people are monitoring your actions—whether they are immediate colleagues, peers whose respect you crave, or a sponsor expecting a return on their investment—tends to increase the pressure when making difficult decisions. Even if a decision maker is not overly concerned about their personal image, the presence of other people can still interfere with good decision making. Although too much agreement can result in groupthink, disagreements can also lead to undesirable outcomes such as vacillation, decision avoidance, compromise decisions, or outright conflict. Interpersonal tensions tend to increase stress levels, reducing effectiveness even more, as we saw in Chapter 7.[20]

Our decision making is also subject to biases in thinking that tend systematically to distort our judgement. In everyday life, these biases can lead us to make trivial or non-trivial (but rarely fatal) mistakes. In extreme environments they can be deadly.

Three particular types of cognitive bias have a direct bearing on the judgement of risk. We all have a tendency to be over-optimistic about the likelihood that risky events will befall us (optimism bias) and to overestimate our competence, relative to others, in dealing with difficult situations (illusory superiority bias). We also tend to judge some risky events to be more likely than others, despite their actually being much less likely, because examples of those events come more easily to mind (availability bias). Together, these cognitive biases can lead people in risky situations to underestimate the real dangers and overestimate their ability to cope, while focusing unduly on relatively unlikely risks.[21]

Optimism bias is generally stronger in adolescence than in adulthood, which might contribute to the greater tendency of adolescents to engage in risky or reckless behaviour.[22] Despite understanding that a particular type of risk exists, such as dying in a vehicle accident, they systematically underestimate the likelihood that it will happen to *them*.

Age and experience appear to have a moderating effect on optimism bias. Research on BASE jumpers, paragliders, and parachutists, for example, suggests that experienced practitioners have a realistic (comparatively pessimistic) view of risks.[23] Individuals with long experience of a risky activity are likely to have had more personal exposure to bad outcomes, making them more realistic about the dangers they face. Being aware of how close you can be to catastrophe is a powerful incentive for thinking through the risks.

Reminding people of their own mortality does not necessarily make them more risk-averse, however. Indeed, it can do the opposite. Although the evidence is inconsistent, some research suggests that individuals who have narrowly avoided death may, under certain circumstances, subsequently expose themselves to more risks. Some risk-taking behaviour, such as smoking, drinking, and taking illicit drugs, can be a defensive reaction to traumatic stress. Another explanation is that when someone has faced the prospect of death, other risks seem less daunting. Having cheated death, the person behaves as if they were immortal.[24]

Although over-optimistic novice practitioners may underestimate the risks of extreme activities, the general public usually overestimates them. Research suggests that experienced practitioners have a more realistic perception of the risks they face than do the general public.[25] One reason may be that most people tend to hear about extreme activities through the media when something has gone wrong. Practitioners, on the other hand, know about the many more occasions on which nothing has gone wrong. The overestimation of exotic risks is an example of availability bias. We tend to believe that a particular type of event, especially a vivid or shocking one, is more likely than is actually the case when instances of it come easily to mind. A typical example is the tendency to believe that being killed by a shark is more likely than being killed by falling aircraft parts, simply because we are more likely to have read about shark attacks.[26]

Illusory superiority—our tendency to believe we are better than most people at a particular task—can result in poor judgement, especially when combined with over-optimism. Studies have shown that most of us think we are better than average in a range of everyday skills, such as driving and problem solving. Research has demonstrated that people such as doctors and professors also fall victim to illusory superiority effects in their professional lives.[27]

Another form of bias that affects decision making is risk compensation (or risk homeostasis). This refers to our tendency to take more risks when we feel safer—for example, because we are using protective equipment. Modern safety equipment makes disasters such as crashing your car more survivable than they were in the past. According to the theory of risk compensation, that very feeling of safety makes us more inclined to take the risks that lead to disaster. Consequently, the overall injury rate may stay the same or even increase.[28] As one skydiver put it, 'the safer skydiving gear becomes, the more chances skydivers will take, in order to keep the fatality rate constant'.[29]

Evidence consistent with the theory of risk compensation has come from studies showing, for example, that drivers who wear a seatbelt are more likely to speed, or that the introduction of childproof safety tops on medicine bottles initially lured some parents into being less careful about keeping bottles out of the way, leading to an increase in child poisoning.[30] Similarly, a study of risks taken by children while cycling or rollerblading found that those who wore safety gear took more risks than those who did not wear safety gear.[31] The theory of risk compensation is controversial, however, and other studies have found little or no evidence for the effect. One problem for researchers is the difficulty of measuring risk-taking behaviour, as distinct from outcome measures such as injury or fatality rates.[32]

As we have seen, environmental and contextual stressors in hard places make people more susceptible to decision biases and ill-calculated risks. An example, well known among climbers, is the phenomenon of 'summit fever', in which the desire to achieve a summit, no matter what the cost, overrides all other considerations. Summit fever is not caused by a single factor, but results instead from a combination of biases made more likely by the stressors encountered in extreme environments.[33]

K2 is the second highest mountain on the planet, and is generally considered to be a harder and more dangerous climb than Everest. One in four of those who attempt it never return. In August 2008, twenty-four climbers were ascending K2 when they became stuck in a queue as they tried to negotiate a steep gully called, appropriately, the Bottleneck. Some turned back, but most continued. They had spent small fortunes getting there, had long dreamed of this day, and they only had one shot. When people have invested heavily in a particular course of action they tend to press on in the hope of getting a return on their investment, even in the face of mounting evidence that to continue would be folly—a cognitive bias known as the 'sunk cost bias'.[34]

Another cognitive bias was also at work that day on K2. In situations of uncertainty, we tend to do what everyone else is doing, even if it makes us uneasy.[35] Climbers on K2, seeing that no one else was turning back, followed the herd, pressing on despite any doubts they might have had.[36]

These highly experienced climbers took ill-judged risks. Complex Himalayan climbs entail long periods of exposure in a thin atmosphere, where you burn calories and dehydrate far faster than you can replenish your reserves. Debilitating altitude sickness can strike even the most robust mountaineer, and exhaustion makes a fall more likely. Where the safety margin is so narrow, climbers must ensure that their actions do not increase the risks any further. A critical part of any high-altitude climbing plan is the turn-around time: the time at which an ascending climber must turn back if they are to survive, regardless of how close they are to the summit. The longer a person stays on a dangerous mountain, the more likely they are to fall, be caught in an avalanche, or develop altitude sickness, exhaustion, or hypothermia.

Having lost valuable time in the Bottleneck, most of the K2 climbers who eventually summited did so very late in the day. They had left themselves no margin for error. Night was falling and, as they attempted to descend, an ice cliff that lay in their path began disintegrating, pelting them with tons of ice, sweeping away their ropes, and obscuring the route. Eleven climbers died.[37]

Expertise

Despite the many cognitive biases and environmental stressors that can interfere with judgement, people in extreme environments usually do make good decisions, which is why they usually survive. Personal characteristics such as emotional stability and low trait anxiety help when making good decisions, as do behaviours such as getting enough sleep. But possibly the single most important factor is expertise—that keenly honed set of skills, knowledge, and wisdom that enables someone to cope with demanding circumstances.

By the late 1980s, applied psychologists were becoming frustrated by the lack of useful theory to explain how people make decisions in real-world situations. Empirical studies of decision making had been carried out for decades, but these had mostly been laboratory studies of undergraduate students. Most researchers chose to control for the potential effects of expertise by ensuring that none of the student participants had any specialist knowledge of the tasks they were to perform. What these experiments gained in control, however, they lost in applicability to real-world decision making.[38]

A few psychologists took a different approach: they were looking at people who regularly used their professional expertise to make important decisions in demanding situations. These psychologists set about investigating decision making in the high-stakes environments of the military, the nuclear, oil, and gas industries, emergency services, and many domains that fall within our definition of extreme environments.[39] Their emerging field of study became known as naturalistic decision making (NDM).

NDM research began with observational studies but later developed to include controlled experiments. At the heart of all these studies is an insistence that experience and expertise are integral to understanding how people solve problems in the real world.[40] Researchers soon uncovered various ways in which expert decision makers differ from novices in high-stakes situations. One of the most important is how experts use their knowledge to gain a deep understanding of a decision situation.

All decisions have two broad elements: understanding the situation in which the decision must be made, and choosing from a range of actions that are appropriate to the situation as understood. In extreme environments, as in everyday life, the first step in decision making is ensuring you know exactly what you are making a decision about. This is usually obvious in routine situations, but under ambiguous and stressful conditions the way in which the decision maker understands the problem is crucial. If you misunderstand the situation it may not matter what decision you make, because it will probably be the wrong one.

Psychologists describe the process of understanding a decision situation as 'situation assessment'. In straightforward conditions, situation assessment involves classifying the problem according to contextual cues and recognizing it as typical of a particular type of situation. In more complex conditions, situation assessment involves more complex sense-making, where the decision maker draws on relevant knowledge and makes inferences to understand the situation.[41]

Once you understand the situation, whether by recognition or reasoning, the choice of action may be obvious, in which case it is simply a matter of retrieving from your memory the most appropriate action for that particular type of situation. However, even in a well-understood situation there may be more than one possible option, and choosing one rather than another may involve trade-offs.

These processes happen constantly in everyday life. Throughout your waking hours, your brain is making sense of the situations you find yourself in, classifying them as a type of situation that you do (or do not) know well, and choosing an action. Most of this goes on below your conscious awareness, where the process of understanding a situation and choosing an action is so straightforward that you are unaware of having even made a decision. Sometimes, however, you have to stop and think about what is happening and what you should do next.

In the domains in which they are proficient, experts differ from non-experts in both the situation assessment stage and the 'action choice' stage of the decision process.[42] First, they are more likely to pay attention to relevant details. Novices are at risk of paying attention to everything, and thereby becoming overwhelmed with information, or missing important cues, or focusing on irrelevant details. Where an experienced mountaineer notices different textures of snow, a novice may see only an expanse of whiteness.

Next, experts are more likely to make correct inferences about what they perceive, and therefore more likely to assess the situation correctly. Even had they paid attention to it, the texture of snow might mean little to a novice, whereas to the expert it signals whether the snow has settled securely and is relatively safe, or is a fresh fall that indicates avalanche danger.

With their greater exposure to their domain of expertise, experts have a broader and richer repertoire of similar situations stored in their long-term memory, against which to match the current situation. Our expert mountaineer may have crossed apparently benign snowfields before and experienced the shock of triggering an avalanche. Or they may have learned from others how to distinguish safe pathways from treacherous ones. This knowledge helps them diagnose their current situation as 'probably safe' or 'potentially dangerous'.

Although we have described the process as though it were linear, the perception and recognition process is in reality bi-directional. Cognitive psychologists refer to 'top down' (knowledge-driven) and 'bottom up' (cue-driven) processing. The top down process draws on your store of knowledge; it tells you what to look for and attend to ('I'd better keep an eye out for a particular type of snow'). The bottom up process involves noticing detailed cues indicating that a particular situation might be present ('I can see the snow has a particular texture: that might mean trouble').

When one of the founders of the NDM movement, Gary Klein, set out to study firefighters' decision making, he was surprised when they told him they did not make decisions, they just acted. When Klein asked them why they took particular actions, they struggled to articulate their reasoning. So he tried another tack. Using an interview method called the Critical Incident Technique, he asked experienced firefighters to talk through what they did in the course of a memorable fire.[43]

When he and his colleagues analysed the interview transcripts, Klein found his answer. These experienced professionals were picking up on subtle patterns of cues relating to the type of fire they were dealing with, how it might spread, and where the danger spots might be. Recognition of these cue patterns then triggered the retrieval of particular action choices from the firefighters' extensive memories of previous fires.

Klein and other NDM researchers started seeing this same process of recognizing cue patterns and retrieving action choices in their research with other professionals in high-stakes environments, including military commanders, pilots, nuclear power station personnel, astronauts, and police officers.[44] Because experienced decision makers encounter the same sorts of situations on a regular basis, the process of pattern recognition and action retrieval becomes automatic. Just as we all take thousands of actions every day without being conscious of making a decision, so experts in a particular domain often appear to act without consciously making a decision, even though many alternative courses of action may be available. That is why Klein's firefighters claimed they did not make decisions.

Klein developed a model of expert decision making, known as the Recognition Primed Decision Model (RPDM), which has become one of the most influential and best-tested theories in NDM research. It explains how experts usually make superior decisions and why experience matters so much.

The model also highlights the importance of training novices how to assess a situation. A person does not necessarily require any deep understanding of a situation to make 'simple match' recognition-primed decisions. Simple rules, of the type 'take this action when you observe these features', may be sufficient in straightforward situations. However, training people to follow simple rules like these tends to produce what one researcher has described as 'surface manifestations of competent behaviour'.[45] Novices may appear to be acting competently in dangerous situations without truly understanding the dangers they face.

The simple decision making process described by Klein's RPDM is not suitable for all situations in which experts make decisions. The pattern of cues may be idiosyncratic, or too subtle, ambiguous, or unfamiliar for even a highly experienced individual to recognize automatically. Furthermore, the action choices may be unclear, or have such potentially serious consequences that an action cannot be taken without further deliberation. In these more demanding circumstances, the process of situation assessment is a more complex version of the top down/bottom up process we described earlier.

When someone is trying to make sense of a complex situation, they may construct stories in their mind to reach an understanding of how the situation arose, its most important features, and, crucially, how it may develop. These stories are built by integrating new information with memories to construct a mental model of the problem situation.[46] (As you may recall from Chapter 8, mental models are ways of organizing information in our minds.)

Expertise is crucial in the process of assessing complex situations. Experts have a better understanding of the antecedents, key features, prognoses, and implications of the situation they are faced with, and therefore construct more detailed, comprehensive, and coherent mental models than novices. They then search their memory for similar situations they have encountered in the past, at which point the choice of action often becomes obvious.

Novices, on the other hand, are less able to construct coherent stories, and tend to spend less time doing so. Lacking the expert's deep understanding, the novice tends to think that everything is important, or that the wrong things are important, or that nothing is important. Because the situation remains ambiguous to the novice, a single choice of action is not necessarily obvious and so they must spend time generating and deliberating among options. Whereas experts spend more time diagnosing the situation and less time choosing a course of action, novices are inclined to focus on possible actions at the expense of situation assessment.

What if the best course of action is not immediately apparent, perhaps because the mental model generated in situation assessment does not map on to a familiar situation, or because the situation is unpredictably dynamic? Compared to novices, experts are better able to generate a feasible course of action despite never having encountered the situation before. In this sort of circumstance, experts engage in mental simulation of a potential course of action before implementing it.[47]

Expertise and experience have their limitations, however. There are some circumstances where even the most skilled practitioner can be caught out. The mountaineer Ed Viesturs asserted that statistics 'didn't apply' to him and that each climb made him 'smarter, faster, stronger, more efficient'.[48] But experience does not always confer the protection that some practitioners believe it does.[49] As they gain experience, many mountaineers challenge themselves with increasingly novel and dangerous climbs. However, the lessons learned on easier climbs do not always work in riskier situations, and some climbs are so inherently dangerous that even extensive experience can alleviate the danger only a little. As the inherent risks get bigger, the difference between moderately experienced and highly experienced practitioners can eventually become marginal (what psychologists call a ceiling effect). The same principle applies in other extreme environments.

Planning and preparation

Many experts would argue that the decisions made before an extreme mission starts are often more important than the decisions made during the mission itself. Planning and preparation make the difference between life and death.

The importance of planning was obvious to Roald Amundsen, who in 1912 beat Captain Scott in the race to be first to the South Pole. When Amundsen returned to his ship, he wrote: 'Victory awaits him who has everything in order. Luck, people call it. Defeat is certain for him who has neglected to take the necessary precautions in time—this is called bad luck'.[50] Some commentators have argued that errors of planning contributed to Scott's failure: had he not skimped on rations for his final, fatal trek, he and his companions might have survived their return journey. Scott had originally planned for a party of four to make the Pole attempt. In the event, he decided that five would go, but did not increase the rations accordingly. Scott and his men were also genuinely unlucky: an unusually fierce blizzard trapped them in their tent, 11 miles from the nearest food cache.[51]

Meticulous planning is essential when exploring the world's most remote places, such as deep caves. Some have likened exploring deep caves to climbing Everest, although without Sherpas to carry the heavy gear. Take, for example, Chevé Cave in Mexico, which stretches almost a kilometre below the surface. The lengthy route down includes sheer drops of 150 metres, waterfalls, crawl spaces, lakes, and huge boulder fields. When water blocks the way it has to be dived through—the most dangerous type of diving. The hardest part, climbing up and out, comes last, when cavers are most exhausted.[52]

Bill Stone, one of the world's most experienced cave divers, stressed the importance of exhaustive planning, which cannot eliminate risk but can at least reduce it to an acceptable level. Speaking in 2004, Stone observed that he had lost sixteen friends to caving accidents over an eighteen-year caving career. The only way to plan, he asserted, is to treat a cave 'as if it is actively out to get you'.[53] He recommended using checklists, rechecking every last detail, and building in redundancy. When you are miles below the surface and your gear fails, you need a backup instantly available. Most important, said Stone, is to be willing to abort. Turning around is difficult, but proceeding when problems have come to light is asking for trouble. His friends were killed not by one single disastrous event but 'by a string of little events'. Stone advocated taking note of every problem, however minor, because in an extreme environment an accumulation of 'little events' can quickly spiral out of control.[54]

By the time Charles Lindbergh took off to make the first solo flight across the Atlantic, he had prepared meticulously, ensuring that every mechanical detail of his aeroplane was correct. He had, however, made one serious error, which was failing to get sufficient sleep the night before. As we described in Chapter 4, Lindbergh had to battle against sleepiness throughout his 33-hour flight and almost died as a result. That seemingly small omission in planning came close to killing him.[55]

Most of us are not very good at planning. Projects usually take longer and encounter more problems than are predicted at the outset. This tendency to be over-optimistic about completing projects on time is referred to by psychologists as the 'planning fallacy'.[56] It has been observed in a range of business, personal, and academic contexts. For instance, most people are over-optimistic about when they will complete their tax return and underestimate how long they will take to complete academic work. The planning fallacy is a notoriously common flaw in large-scale IT and infrastructure projects.[57]

The planning fallacy is not simply a result of ignorance. People usually have access to information about how long it takes to complete similar tasks, but ignore this when making their own judgement. We all tend to focus on the particular case at hand, while ignoring broader contextual information. When we do think of similar cases, we tend to remember those projects that were successful and disregard evidence of when we failed to complete tasks on time. We find it hard to imagine many different possible outcomes for a project, and tend to think of positive outcomes more readily than negative ones. Consequently, we often struggle to contemplate anything other than a happy ending.[58]

One method to help planners think about potential bad endings is the 'crystal ball' technique. You make a prediction—for instance, about how long a particular step might take—and then imagine a crystal ball telling you that your prediction is wrong. This forces you to explain why your prediction might be wrong and come up with an alternative prediction. The crystal ball again tells you that your prediction is wrong, and so on. If applied thoroughly, the crystal ball technique forces you to think of bad outcomes as well as good ones.[59]

Another method of avoiding the planning fallacy is systematically to document all the possible ways in which a project might go wrong, and work out what to do in every case. For complex expeditions, such as space travel, this takes time and expertise. This is why space missions take years of planning and preparation, most of it focused, as astronaut Chris Hadfield put it, on 'rehearsing for catastrophe'.[60]

Becoming expert

The evidence is clear that surviving and thriving in hard places requires expertise. But how does someone become an expert? The skills and knowledge required to perform well in extreme environments can be acquired in three ways: through deliberate practice of core skills, through formal training, and through repeated exposure to challenging situations (i.e. experience).

Deliberate practice includes the development of physical abilities and tolerance to physical stressors. As a boy, Roald Amundsen skied and slept with his window open during the freezing Norwegian winter, 'increasing my skill . . . and hardening my muscles for the coming great adventure'.[61] Serious climbers practise weight training to develop their upper body strength, and those who tackle the highest mountains ensure they are in peak physical fitness. An analysis of deaths on Everest found that many climbers on commercial expeditions had failed to take physical preparation seriously, putting themselves and their fellow climbers at risk.[62]

To attain exceptional levels of performance, whether in extreme activities or everyday life, you cannot just rely on natural talent. To excel at anything takes many hours of dedicated practice.[63] Practice involves repetition, such that successful actions are reinforced. Repetition only gets you to 'good', however. To become excellent, your practice sessions must involve effort. As you get better, they should become more challenging so that you continue improving. Good feedback is necessary too. Reflecting on what went well in each session helps to consolidate improvements. But the most important feedback is from failure. Only by failing do you really become conscious of your errors and take steps to correct them. Without accurate feedback, you may become experienced but incompetent.[64]

Not everyone can achieve exceptional performance, and there are reasons why most of us do not become experts in whatever we turn our minds to. One reason is that developing expertise requires material and mental resources. As well as the requisite physical attributes, those seeking to excel need time to train and time to travel, which probably means spending time away from family, friends, and job. They also need money: for travel, equipment, tuition, and simply to live.

Free-divers compete against themselves and others to reach ever greater depths without the benefit of breathing apparatus. They could practise holding their breath in a local swimming pool, but they must dive deep to develop the skills needed to break records and, unless they happen to live somewhere suitable, they will have to travel. Mountaineers practise on climbing walls, but they must also climb outdoors to prepare for more challenging feats. Exploring extreme environments is seldom cheap.

Deliberate practice is hard work and not always particularly enjoyable. Those who are driven to achieve a goal, whether to run a marathon, pass an exam, or reach the top of Everest, must be both determined and persistent. Psychologists have dubbed this trait 'grit': the demonstration of 'effort and interest over years despite failure, adversity, and plateaus in progress'.[65] Grit is a measure of an individual's propensity to work hard, regardless of how clever they are. And hard work pays off: compared to less gritty people with similar IQ scores, individuals who score highly on measures of grittiness tend to have higher levels of educational achievement, and gritty recruits are more likely to complete gruelling military courses.[66]

Single-mindedness is important too. In one study, researchers compared an elite international group of swimmers, many of whom were Olympic Gold medal winners, with a group of 'sub-elite' swimmers who were at the top of their national ranking. The researchers found no clear difference between the two groups in the amount they trained or the social and material support they enjoyed. The only apparent difference was that, from an early age, the elite swimmers had expressed their intention to be the best, and had maintained this single-minded determination throughout their career.[67] This is not to say that the qualities you are born with are irrelevant: some individuals are physically more suited to particular activities and, with practice, their initial advantage may grow stronger.[68] As one elite swimming coach commented, 'hard work beats talent, until talent decides to work hard'.[69]

Another way in which individuals develop expertise is through formal training. The training for extreme environments, as for more conventional environments, is generally a combination of 'book learning' and experiential learning. While the former can equip aspiring adventurers with essential background knowledge, experiential learning helps them apply that learning to the situations they are likely to encounter. Supervised experience of dangerous situations helps people prepare for the emotional as well as physical challenges of extremes.[70]

Experiential learning does not have to take place in the real extreme environment, even if that were possible. What matters is that the training environment makes similar physical, cognitive, and emotional demands. This principle was illustrated by a 1990s study, in which Brazilian armed forces officers on ships heading for Antarctica rated their anxiety levels at various points during the voyages. The ships encountered bad weather and were sometimes left without power, drifting in hazardous waters. Despite this, and contrary to expectations, the crews reported relatively low levels of anxiety, similar to those shown by people in non-stressful conditions. Their relaxed stance persisted even in conditions that most people would consider highly stressful. The officers had previously undergone extensive training in dangerous situations, including in the jungle and on remote islands. Although these environments were not direct analogues for the Antarctic Ocean, the officers had developed general coping mechanisms that helped them to manage stress in different hazardous situations.[71]

The transfer of expertise from one situation to another is one reason why people destined for extreme environments prepare in simulations. For many years, simulations have been a mainstay of training for environments that are too dangerous, expensive, or impractical for novices. Simulated and analogue environments are used extensively by NASA to prepare astronauts for space travel and are widely employed in other extreme occupations, including the military.[72]

The wartime US intelligence agency the OSS (Office of Strategic Services, the forerunner to the CIA), for example, used a mix of physical and psychological 'toughening up' in its officer training. Trainees were going to be deployed on dangerous missions, in which they would spend months behind enemy lines gathering intelligence and carrying out acts of sabotage. In preparation, they were put through their paces in alarming simulations. Mocked-up houses, where 'targets with photographs of uniformed German soldiers jumped out from every corner and every window and doorway', were designed to 'test the moral fiber of the student and develop his courage and capacity for self-control'.[73]

Simulations are an important means of gaining relevant experience for extreme environments. The closer the simulation is to reality, the more readily the lessons are transferable. However, there is no substitute for accumulating experience in real situations. People who are determined to excel in extreme environments therefore repeatedly expose themselves to increasingly challenging situations, learning lessons and building resilience as they go.

This brings us to the final ingredient in building expertise: continuous reflection on lessons learned. After completing a dive, or a climb, or some other extreme activity, experts consciously assess how they performed and think about how they could do it better next time. The mountaineer Joe Simpson wrote that analysing what he had done correctly or incorrectly was as important as being fit or talented.[74] Those individuals who learn deeper lessons from their experience may, over time, develop wisdom.

Scholars have debated the nature of wisdom for thousands of years, although without arriving at a universally accepted definition. The psychologists who have contributed more recently to this debate have defined wisdom in various ways, including 'expertise in the conduct and meaning of life', 'the ability to act critically or practically in any given situation', and 'understanding how to apply concepts from one domain to new situations or problems'.[75] Inevitably, they have created various rating scales to measure it.[76]

Psychological descriptions of wisdom often refer to the ability to cope well with ambiguity and uncertainty—a crucial ability in extreme environments, as we have seen. Knowledge and cognitive skills are necessary ingredients of wisdom but they are not sufficient. You can be intelligent and know a lot without being wise. Age and experience are also associated with wisdom but, again, they are not sufficient. Wisdom does not grow automatically with age and it remains a relatively rare attribute even among those with decades of experience.[77]

Despite its popular portrayal as an almost mystical quality, research suggests that wisdom is learned. More specifically, it is the way in which people reflect on and learn lessons from experiences, especially challenging, unusual, or even traumatic experiences that determines their performance on the main dimensions of wisdom.[78] It appears that the path to wisdom lies in learning the right lessons from hard experiences.

To conclude, we have seen that surviving and thriving in extreme environments requires good judgement, good decision making, careful planning, and thorough preparation. To make good decisions, whether in extremes or in everyday life, you need to understand the decision situation correctly and then choose the appropriate actions. All decision making is subject to cognitive biases. Expertise mitigates, but does not completely eliminate, such bias. Experience, deliberate practice, and formal training help to develop the expertise required to make good decisions. Becoming an expert is hard work, but it can be a source of great satisfaction.

10

Focus

Be master of your petty annoyances and conserve your energies for the big,
worthwhile things. It isn't the mountain ahead that wears you out—it's the
grain of sand in your shoe.

Attributed to Robert Service (1874–1958)

Expertise is critical, as we saw in Chapter 9, but distractions can hamper even the best-trained expert. In this chapter we consider a fundamental skill that underpins good decision making. It also contributes more widely to success and well-being, both in extreme environments and everyday life. That skill is focus—the ability to pay attention to the right things, and ignore all distractions, for as long as it takes.

Most extreme activities are impossible without a high degree of focus. Consider free solo climbing, for example. This entails climbing a sheer wall of rock with no rope. If you lose your grip, you fall and die. At the time of writing, Alex Honnold is widely regarded as the world's leading free solo climber. He has made some remarkable climbs, including the colossal walls of the 2,000-foot Half Dome and the 3,000-foot El Capitan in Yosemite National Park. The concentration required is extraordinary. Half Dome took nearly 3 hours to climb the first time he scaled it, and Honnold was one slip away from death for every second.[1] Without strong focus, such feats would be impossible.

The intensity of experience in some extreme environments has a way of commanding attention, even in someone who may not be consciously trying to concentrate. As one scholar noted, 'the prospect of death produces a powerful form of attention focussing'.[2] For some adventurers, that intense focus is one of the most rewarding aspects of what they do. The psychologist and mountaineer James Lester wrote that 'focusing of attention and whole-heartedness of purpose . . . and their consequence of vividness, intensity, and transcendence are not easy to achieve in everyday life, and may be one of the deeper things that mountaineering is all about'.[3]

The ability to control attention and ignore distractions is a crucial but much underestimated skill in everyday life as well. Without it, our perform-ance suffers and we become prey to chronic dissatisfaction. Focus is good for everyone's well-being. The people who survive and thrive in hard places are often experts in focus, even if they do not know it. What can we learn from them?

Paying attention

According to an old adage, the only way to eat an elephant is one bite at a time. The metaphor describes a simple technique used by many people in demanding situations: take things one small step at a time, focusing tightly on the next bite-sized action rather than dwelling on the entire daunting mission.

Apsley Cherry-Garrard did this when he and his two companions made their nightmarish Winter Journey, slogging across the Antarctic wastes in pitch darkness. He kept himself going by refusing to think of the past or the future and living only for 'the job of the moment'. To help himself focus, Cherry-Garrard invented a verbal 'formula' that he repeated to himself continually. His formula (or mantra, as some might now call it) was: 'You've got it in the neck—stick it—stick it—you've got it in the neck—stick it—stick it . . .' and so on.[4]

Joe Simpson used a similar strategy after breaking his leg on an Andean mountain and being left for dead in a crevasse. Dehydrated, exhausted, and dragging his shattered leg, Simpson made it down the treacherous slopes by dividing the immense task of surviving into much smaller subtasks that he could focus on. He would set himself the goal of dragging himself for one

20-minute period, then another, then another, and so on. Over the course of three days Simpson crawled an agonizing 5 miles to safety.[5]

The Danish Sirius Patrol is a military unit that patrols the north-eastern edge of Greenland for months at a time in conditions of isolation, extreme cold, and 24-hour darkness. Psychologists who studied the unit in 2013 reported that officers' most common coping strategy was to remind themselves to 'take it one day at a time'.[6]

Focusing is not easy. When the mind is not fully absorbed by a demanding activity, it has a natural tendency to wander. Instead of focusing on the here-and-now, it drifts off to inner musings, feelings, or fantasies, flitting readily from thought to thought. Our attention shifts away from current sensory input and turns to thoughts and feelings that are relevant to the self. This state, which can be quite pleasant, is variously referred to as mind-wandering, daydreaming, absent-mindedness, or zoning out.[7] Mind-wandering is a universal human experience. Empirical evidence suggests that between 15 and 50 per cent of our waking time is spent in this state. The mind is more likely to wander when working (or short-term) memory is not fully occupied with a task.[8]

As you would expect, mind-wandering weakens cognitive performance, especially the ability to perform complex tasks that make large demands on working memory. Lapses in attention, thinking errors, and forgetfulness become more likely, with potentially serious consequences for anyone performing safety-critical tasks. Instances of mind-wandering have resulted in pilots failing to lower an aeroplane's landing gear and surgeons leaving instruments inside their patients.[9] Lapses in attention also cause us to miss important pieces of information, which impairs our ability to make sense of complex situations. In one experimental study, for example, people whose minds wandered were less able to understand the complicated plot of a detective story.[10]

Although mind-wandering is hazardous when performing safety-critical tasks, it does have some benefits. Most of the thoughts that arise during mind-wandering are about the self and the future. This has led some psychologists to suggest that mind-wandering may help people to anticipate and plan for personally relevant events.[11] Sometimes the wandering mind may stumble across interesting new ideas or associations, and there is some evidence to link it with heightened creativity.[12] Even so, you would not want your mind to be wandering if you were engaged in a potentially fatal activity.

Extreme environments include many potential distractions. One of the most potent is fear. As one mountaineer wrote, 'the trickiest moves on any climb are the mental ones: the psychological gymnastics that keep terror in check'.[13] Divers also know that their activity is fundamentally a mind game. If something goes wrong deep underwater, the ability to focus the mind and make the right decisions will be critical to survival. A diver who becomes disorientated in an underwater cave, or whose equipment malfunctions, has a much better chance of surviving if they can maintain a cool focus on the immediate problem.

The ability to remain focused under water can also be impaired by relatively minor distractions or upsets that occur before the dive, such as delays, equipment problems, or arguments with colleagues. For that reason, some experienced divers make it a rule to postpone a challenging dive if, say, three unforeseen problems occur before the dive, regardless of whether those problems have been fixed. Other divers have discovered that silently repeating a word, or mantra, helps them enter a calm and focused state.[14]

Noise and vibration are common causes of distraction in closed environments such as submarines, spacecraft, and other artificial habitats. The occupants usually habituate to continuous noises, like the drone of a generator, and some even find them soothing. (Habituation can be a problem, however, when an alarm goes off so regularly that people start ignoring it, as happened on the Mir space station.[15]) Alternatively, crews may experience persistent tension as they listen 'unconsciously' for sounds that indicate mechanical problems.[16]

The effect of noise on attention and hence cognitive performance depends on the type of noise and the type of task. People engaged in tasks that involve processing words are more easily disrupted by voices than by white noise. A noise that changes in pitch or intensity is more distracting than a monotonous sound. We automatically pay attention to stimuli that signal danger, so frequent changes of background noises and intermittent blasts of noise are more likely to cause stress, especially if they are unpredictable and uncontrollable. The after-effects of stressful noise can last for days.[17]

Individuals vary considerably in the ease with which they are distracted by noise. This variation can be measured with a psychometric tool called the Weinstein Noise Sensitivity Scale (WNSS). If you strongly agree with statements such as 'at movies, whispering and crinkling candy wrappers disturb me', and strongly disagree with statements such as 'I am good at concentrating no matter what is going on around me', then you are likely to score highly on the WNSS. Individuals with high WNSS scores are less able to concentrate on cognitive tests in busy environments, and tend to do less well at school.[18] Personality plays a role. Studies have found that noise has a negative impact on the cognitive performance of introverts and anxious people, whereas extraverts may actually perform better in noisy conditions.[19]

Illness, which is common enough in extreme environments, can be a powerful disruptor of focus. Even minor infections are accompanied by measurable changes in memory and other aspects of cognitive performance. More severe infections are typically accompanied by lethargy, mood depression, and general malaise. These psychological and behavioural responses to infection have evolved to assist the recovery process and keep the sick individual out of harm's way. The immunological response to an infection triggers the release of chemical messengers that act on the central nervous system to induce drowsiness, fatigue, an increase in slow-wave sleep, and loss of appetite.[20]

A depressed mood, such as might arise in extreme environments during periods of adversity or loneliness, is another potential cause of poor focus. Even a mildly depressed mood makes the mind more likely to wander, as thoughts drift towards inner concerns. Experimentally inducing a low mood, by showing people disturbing video clips, reduces their ability to concentrate on cognitive tasks, impairing their performance. People in a depressed mood also find it harder to recover from lapses in attention and refocus on a task. Conversely, exposing people to mood-enhancing video clips improves their focus and cognitive performance.[21]

Willpower

When Apsley Cherry-Garrard was asked to describe the essential characteristics of a polar explorer, he said: 'It is a matter of mind rather than body', adding that 'it is not the strength of body but rather strength of will which carries a man farthest'.[22] Willpower—the ability to focus effort and ignore distracting urges—is an essential ingredient for success in extreme activities and in everyday life.

The nineteenth-century psychologist William James regarded willpower as the ability to control attention and thereby ignore 'fiery passion's grasp'. More recently, psychologists have variously defined willpower as: the deliberate control of powerful impulses and appetites, the determination to complete a chosen or assigned task and ignore competing tendencies, and the ability to override one's urges and emotions. A common thread running through most definitions of willpower is self-control, and some psychologists regard the two as synonymous. Empirical evidence shows that individuals who score highly on a psychological measure of willpower are more likely to do what they said they would do, such as preparing for an exam. They also tend to score highly on the personality trait of conscientiousness.[23]

One well-known measure of willpower is delay of gratification. In the 1960s Walter Mischel and colleagues at Stanford University tested the willpower of young children by making them choose between instant gratification, in the form of one delicious marshmallow now, and a larger but delayed prize of two marshmallows later. The children who exercised self-control and won the bigger prize grew up to be healthier and wealthier. Children's scores on delay of gratification were remarkably predictive of positive outcomes later in life, including higher educational achievement, better emotional coping in adolescence, greater self-worth, greater resistance to stress, lower likelihood of using addictive drugs, and lower risk of becoming overweight.[24] Other research has shown that individuals with strong willpower or self-control also do better in terms of mental health, personal relationships, popularity, and coping skills, while being less likely to engage in aggressive behaviour or criminality.[25]

If exercising willpower can seem tiring, that's because it is. Experiments in the 1990s showed that people's ability to complete a task requiring effort and persistence was impaired if they had recently been forced to exercise willpower—for example, by resisting the temptation to eat freshly baked biscuits. Participants who had not been tempted with biscuits performed better. These and other findings suggest that willpower is a finite resource, which is depleted by acts of self-control and needs time to replenish.[26]

More recent evidence suggests that one of the resources that is depleted by the exercise of willpower is glucose. Laboratory experiments have shown that even relatively small acts of willpower are followed by measurable short-term reductions in the supply of glucose to the brain, with consequential dips in the ability to exercise self-control.[27] Willpower is restored by consuming more glucose. In one experiment, participants had to make a difficult choice between two options while ignoring an apparently irrelevant decoy option. The participants were better able to ignore the distracting decoy and make the right choice if they had just drunk lemonade sweetened with sugar, but not if the lemonade was sweetened with an artificial sweetener which had no effect on blood glucose.[28]

We can learn to strengthen our willpower and improve our ability to resist urges. One proven strategy is to direct one's attention away from the object of temptation. In the Stanford experiments on delay of gratification, the children who demonstrated the strongest willpower were often those who deliberately averted their gaze away from the tempting marshmallow. A more sophisticated version of this technique involves cognitive reappraisal, which means thinking about the tempting object in a way that makes it seem less attractive. For example, you might pretend that the marshmallow is a ball of cotton wool. What both techniques have in common is focus—the ability to attend to the right things and ignore distractions. Psychologists have suggested other techniques for bolstering willpower, including deliberately avoiding known sources of temptation, ensuring that any lapse has a price, getting other people to impose controls, and, of course, keeping one's blood glucose levels topped up.[29]

Flow

Focus sometimes manifests itself in an even more rewarding form. In extreme environments, and in everyday life, people sometimes experience a positive psychological state known as flow, in which they become utterly absorbed by the task they are performing and lose sense of time. Flow occurs when a person is fully immersed in a demanding activity that stretches, but does not exceed, their level of skill. It is the antithesis of existential boredom, which we discuss later in this chapter.

The concept of flow was originally developed by the psychologist Mihaly Csikszentmihalyi. At the time, he was investigating why people are motivated to engage in leisure activities that have no obvious material benefits or pay-off. Based on a large body of interview data, Csikszentmihalyi identified a rewarding and intense mental state of engagement, which he called flow. He described it as 'a state in which people are so involved in an activity that nothing else seems to matter; the experience itself is so enjoyable that people will do it even at a great cost, for the sheer sake of doing it'.[30] Flow is measured using self-report questionnaires and interviews.

A person who is in a state of flow experiences a range of positive sensations, including intense concentration, feeling in control, loss of awareness of time, loss of self-consciousness, and clarity of goals. The task they are performing feels almost automatic, as though their thoughts and actions have melded together.[31] Studies have shown that, for many activities, the experience of flow is correlated with the level of performance at the task. For instance, athletes who experience flow tend to perform better.[32]

Flow can be elicited by risky activities.[33] A British Army bomb disposal operator wrote of entering a flow-like state while defusing a bomb:

> I've been on task for almost three hours...It only feels like ten minutes. It's what we call operator time. Another world where outside sounds become muted and operators are aware only of the sounds of their own breathing and the drumming of their hearts.[34]

Divers and climbers have similarly described a state of complete absorption and focus that has the hallmarks of flow. Some talk of being 'in the moment' or 'in the zone' and many refer to feeling more intensely alive.[35]

Flow is intrinsically rewarding and therefore tends to reinforce the activities that elicit it. For example, a self-report study of fast motorbike riders found that they experienced the crucial elements of flow while driving at speed. This satisfying feeling reinforced their behaviour and made them more likely to speed in future.[36]

Fortunately, you do not have to engage in dangerous activities to experience the rewards of flow. In fact, flow is most commonly experienced during relatively mundane activities such as reading, writing, playing sport, dancing, making music, or playing computer games. Some games and websites are specifically designed to maximize this effect by providing the user with sufficient challenge to absorb their attention.[37]

The environment in which flow is experienced most often is the workplace. Flow can help to make the daily grind more motivating, although it appears to occur only spasmodically. One study of employees found that individuals' experiences of flow varied randomly over time, suggesting that flow is chaotic and unpredictable. People with relatively complex and demanding jobs, including teachers, surgeons, musicians, social workers, artisans, soldiers, and academics, are more likely to describe their jobs as providing flow, compared to junior office workers and those with routine blue-collar jobs. Nonetheless, almost any job or activity is capable of offering opportunities for flow, provided it is approached with the right attitude, which means becoming fully engaged with the tasks and incorporating personal challenges to make them more stimulating.[38]

Meditation

Until his death in 2009, Slovenian climber Tomaž Humar was one of the world's leading mountaineers. He completed one extraordinary climb after another, on some of the world's deadliest routes. Humar claimed that his secret weapon was meditation. In the hours before each summit attempt, he would find a quiet spot and meditate, concentrating on the forthcoming climb. When things went wrong, such as when he was stranded for a night without shelter at 25,600 feet on a Himalayan mountain, he claimed that meditation helped him control not just his fear but his physiology, enabling him to survive an apparently impossible situation.[39]

Meditation is probably the simplest and best way of cultivating focus. Banish any suspicions you might have about it being new age bunk. As we shall see, there is good scientific evidence that meditation is an effective way to improve the skill of controlling attention and ignoring distractions. It also offers a range of other practical benefits that are relevant both in extreme environments and everyday life.

The term meditation covers a range of techniques that look superficially different but share a common feature: controlling attention. One leading scientists in the field has defined meditation as 'the intentional self-regulation of attention from moment to moment'.[40] It is not the same as relaxing.

Meditation comes in many different forms. Some of these variants involve thick overlays of ritual and beliefs, which have given meditation a misleading aura of quasi-religious flummery. Becoming good at meditation takes time and practice, but it does not require the acceptance of supernatural beliefs. The core reality is relatively straightforward and reassuringly evidence-based.

Stripped to its essentials, meditation techniques are of two basic types. The first and simplest is concentration meditation, in which the meditator concentrates their attention on a single point or object, which may be an image, an imagined sound, or the physical sensation of their own breathing. The second type is mindfulness meditation, which involves keeping your attention rooted in the present moment and observing your thoughts and sensations from a perspective of neutral detachment. The state of mindfulness has been defined as 'paying attention in a particular way, on purpose, in the present moment, and nonjudgmentally'.[41]

What all forms of meditation have in common is learning to control attention. And there is clear empirical evidence that it works. In one typical study, people who were new to meditation learned, over an eight-week period, a form of concentration meditation that involved focusing on their breathing. They subsequently performed better than control participants on psychological measures of attention. In another experiment, participants underwent three months of meditation training. As the training progressed, their ability to focus their attention in a listening task became measurably better and more consistent. Meditation significantly improved their ability to concentrate and ignore distractions.[42]

An extensive array of published evidence shows that focus is a skill that improves with training, and that meditation is an effective form of training.[43] Most of this evidence points to the beneficial effects of practising meditation over long periods. However, even a short burst of training can be beneficial. Experimental studies have shown that just a few days of meditation training can produce measurable improvements in mood and cognitive functioning, including memory and visual processing.[44]

The improvements in focus that result from practising meditation are accompanied by changes in brain activity, some of which are linked to the control of attention.[45] For example, one study found that expert meditators performed better and were less easily distracted than novices when concentrating on a small dot on a computer screen. Brain imaging revealed that the expert meditators had higher levels of activity in areas of the brain associated with sustained attention. When exposed to distracting sounds, the expert meditators showed less activity in brain areas associated with daydreaming, discursive thoughts, and emotions, compared to novices, and more activity in areas associated with attention.[46] The long-term practice of meditation may be accompanied by physical changes in brain structure, including a significant thickening in brain regions associated with attention and sensory processing.[47]

The people who engage in extreme activities might not consciously practise meditation, but their mental techniques have much in common with it. As we saw earlier, a standard way of coping with huge and arduous tasks is to narrow the focus of attention down to the here-and-now, living only in the moment. Extreme activities such as rock climbing require intense and sustained concentration, in which all extraneous thoughts are excluded.

Many extreme environments not only require focus but also provide the conditions that foster it, such as solitude and freedom from mundane distractions.[48] The natural environments in which most extreme activities occur may also play a role in helping people focus attention. Research suggests that viewing natural scenes helps us to recover from the mental fatigue caused by long periods of concentration. In one experiment, people carried out mundane tasks requiring sustained attention until they became mentally fatigued and their performance declined. They then viewed pictures for less than 10 minutes before returning to the tasks. Participants who viewed pictures of natural landscapes, such as orchards, rivers, and mountains, were significantly more accurate in their second set of tasks, compared to people who had viewed urban landscapes or geometric shapes. These and other results suggest that simply looking at natural landscapes can reinforce our capacity to focus.[49]

Boredom

A person who is unable to control their attention is more likely to suffer from a mental state that is the antithesis of flow. That state is existential boredom, an unpleasant emotional condition with unpleasant side effects. Some individuals are driven to seek extreme experiences because they are averse to the tedium of normal life. Paradoxically, as we saw in Chapter 5, extreme environments often expose those same people to extended periods of monotony, a breeding ground for boredom. Boredom can therefore be both a cause and a consequence of extreme activities. As we shall see, it is also a state that has pervasive but widely underestimated effects in everyday life.

What is boredom? The eighteenth-century philosopher Immanuel Kant described it as an unpleasant mental state in which we crave new forms of stimulation and pleasure. Psychologists who have studied the phenomenon agree that boredom is a specific emotion in its own right, rather than an amalgam of other emotions. According to one psychological definition, boredom is an unpleasant, transient affective state in which the individual feels a pervasive lack of interest in and difficulty concentrating on the current activity ... [and] feels that it takes conscious effort to maintain or return attention to that activity.[50]

Crucially, psychologists and philosophers have highlighted a distinction between two fundamentally different forms of boredom. Situational boredom is a temporary and relatively benign state that results from a lack of engaging or pleasurable stimulation from the environment. In other words, it is having nothing to do. Situational boredom is easily remedied by doing something different. In contrast, existential boredom is a deep-seated and potentially intractable malaise.[51]

Existential boredom is an emotional state characterized by a persistent lack of desire for *any* stimulation, no matter how potentially interesting or pleasurable. It has been described as the aversive experience of wanting, but being unable, to engage in any satisfying activity. For someone in the grip of existential boredom, no activity seems desirable any more. They feel unchallenged and restless, in sharp contrast to the rewarding state of absorption that characterizes flow.[52]

A range of empirical evidence suggests that poor focus is a primary cause of existential boredom. More than a century ago, William James observed that one of the prominent features of boredom is the tendency of the mind to wander. He also noticed that individuals who are easily bored find it almost impossible to prevent this: the harder the bored person tries to stop their mind wandering, the more it wanders. Since then, research has demonstrated a robust association between proneness to boredom and a tendency to be easily distracted.[53] People who are easily bored find it almost impossible to prevent their mind from wandering. They tend to interpret their mind-wandering as a sign of boredom, creating a vicious cycle.[54]

The bored person is unable to concentrate on the task in front of them, which makes them perceive the task as unsatisfying. For instance, one study found that individuals who were more easily distracted by non-task-related thoughts were more bored and less satisfied with what they were doing.[55]

Bored people may be conscious of their inability to focus, but they tend to blame this on the inadequacy of the task or the surrounding environment, rather than their own lack of skill. Either way, they feel unhappy.

Even subtle distractions are capable of making us feel bored. In another experiment, participants were asked to remember the content of a moderately interesting article that was read out to them. While they were listening to the article, a television in an adjacent room was playing an irrelevant programme at one of three volume levels (muted, barely noticeable, or loud). The participants who were exposed to the barely noticeable background noise were not consciously aware of being distracted. Nevertheless, they reported feeling higher levels of boredom and found the task less interesting than those who had been exposed to no noise.[56]

Psychologists have investigated the underlying characteristics of boredom and its accompanying features using specially designed psychometric tools such as the Boredom Proneness Scale (BPS).[57] Their research has confirmed that individuals differ considerably and consistently in their propensity for boredom. More significantly, it has revealed that boredom is much more than just a trivial annoyance.

Extensive evidence shows that boredom-prone individuals are more inclined, on average, to exhibit a range of harmful behaviours, including smoking, binge drinking, overeating, criminality, excessive gambling, and the use of illicit recreational drugs.[58] Among young people, susceptibility to boredom is linked to poorer academic performance and a higher risk of truancy and dropout.[59] In adults, higher levels of boredom proneness are associated with higher levels of apathy, job dissatisfaction, absenteeism, and turnover, along with less loyalty to their employer.[60]

Boredom-prone individuals have slower reaction times and perform worse at tasks requiring vigilance—a reflection of their poor ability to control attention. These characteristics are obviously undesirable in extreme environments. And because boredom-prone individuals find it hard to focus on any one activity, they are much less likely to experience the rewarding state of flow.[61]

Boredom can also have corrosive effects on social relationships. According to the evidence, boredom-prone individuals are typically less agreeable, less organized, and more willing to manipulate others. They are also inclined to dislike other people whom they perceive to be dull or uninteresting.[62] These characteristics can make them awkward companions, especially in confined or extreme environments such as polar stations. A US military study found that susceptibility to boredom undermined the cohesiveness, and consequently the effectiveness, of teams deployed on long missions.[63]

Perhaps not surprisingly, in view of all these correlations with unhealthy behaviour, people who are easily bored have poorer mental and physical health, on average, than those who are less easily bored. High scores for boredom proneness are correlated with higher levels of negative emotions, including anxiety, depression, anger, and hostility, and a higher incidence of self-reported mental and physical health problems.[64]

The relationship between boredom and poor health may run even deeper. In a famous long-term health study of thousands of British civil servants, which started in the 1980s, participants answered questions about how they felt, including how bored they had been over the preceding month. More than twenty years later, researchers found that those who had reported feeling very bored in the 1980s were more likely to have died in the intervening years. The results indicated that boredom might have been a proxy for other health-related factors, as it was also associated with poorer self-rated health and lower levels of physical activity. As the researchers concluded, there may well be some truth in the expression 'bored to death'.[65]

The ability to fend off existential boredom is a valuable coping skill, both in extreme environments and everyday life. Given the links between boredom and poor focus, an obvious strategy for reducing susceptibility to boredom would be to cultivate the control of attention. As we saw earlier, one of the best ways of improving our ability to focus is through the practice of meditation. The admittedly limited empirical evidence available suggests that this strategy does help to protect against boredom, in addition to its many other benefits.[66]

In sum, then, coping in extreme environments requires the ability to focus attention on the right things for sustained periods and ignore distractions. Maintaining focus on a challenging activity can be beneficial to psychological health. Focusing attention is not easy, however: the mind naturally tends to wander and we are easily distracted, including by our own thoughts.

Most of us could benefit from strengthening our own ability to focus, not least because poor focus makes us more susceptible to existential boredom, which can be highly unpleasant. Improving focus has the added benefit of making it easier to experience the rewarding state of flow. The ability to control attention can be improved through simple techniques like meditation.

11

Resilience

Sweet are the uses of adversity.   William Shakespeare, *As You Like It* (1599)

M ost people who venture into extreme environments cope remarkably well with the physical and psychological hardships. Their stress levels decrease over time and they often find later that overcoming adversity has brought lasting positive consequences.[1]

The capacity to deal well with pressure is known as resilience. It is the ability to maintain a stable equilibrium in the face of adversity. Resilience is obviously a valuable quality in any setting, whether in extreme environments or everyday life.

In previous chapters, we explored some of the factors that influence resilience. The abilities to face fear, endure hardship, and maintain focus are all characteristics of resilient people. Having supportive relationships, developing expertise, being well prepared, and getting enough sleep are also important contributors to the development and maintenance of resilience. In this chapter we explore the origins of resilience and examine research suggesting that it is a quality we can all cultivate. As we shall see, one of the biggest contributors to resilience is a personal history of coping successfully with previous stressful events.

Being resilient

Resilience is more than just the absence of pathological symptoms: it is a positive quality that equips us to handle unforeseen pressures. The American

Psychological Association defines resilience as 'the process of adapting well in the face of adversity, trauma, tragedy, threats, or significant sources of stress [and] "bouncing back" from difficult experiences'.[2]

Resilient people have an approach to life that is characterized by realistic optimism, self-confidence, a sense of humour, the ability to stay focused under pressure, not being easily defeated by failure, and finding meaning even in negative experiences. They often have a track record of dealing successfully with stressful situations. They also tend to be happier and more successful than those who are less resilient.[3]

Psychologists have investigated resilience in various ways. One approach involves interviewing people who have survived extraordinary hardships, such as former prisoners of war, members of elite military units, or civilians who have overcome traumatic experiences.[4] Another approach involves making psychological assessments of individuals and relating these measures to other aspects of their lives. This often involves measuring resilience using a specially designed self-rating questionnaire such as the Connor–Davidson Resilience Scale.[5] Other research has explored traits that are closely related to resilience, such as hardiness and mental toughness.

Hardiness is a personality characteristic with three main elements: a commitment to seeing life as meaningful and interesting; a belief that one can influence events; and a tendency to see all life experiences, both good and bad, as opportunities to learn and develop.[6] Hardy people are open to change and tend to interpret stressful experiences as a normal aspect of life. Compared to less hardy people, they tend to enjoy better mental and physical health, and are less susceptible to boredom.[7]

Research on armed forces personnel has shown that hardiness provides some protection against the adverse psychological effects of stressful combat. For instance, a study of US military reservists who had recently returned from active service in the 1991 Gulf War found fewer symptoms of stress in individuals who scored higher on a psychological measure of hardiness. The protective effect of hardiness was greatest among those who had experienced the most stressful conditions.[8]

Mental toughness is another concept related to resilience. People who score highly on measures of mental toughness are persistent, focused, and confident in their abilities. They are better able to withstand pressure, cope with anxiety, and endure hardship and pain. Research exploring the personality traits associated with mental toughness has found that mentally tough people tend to be self-disciplined, responsive to new sensations, and not prone to worrying.[9]

The concept of mental toughness has a particular application in sports psychology. Extensive evidence shows that measures of mental toughness are correlated with success in sports. Athletes with high scores for mental toughness are more likely to experience the rewarding psychological state of flow, which we described in the previous chapter.[10] This may be because their self-confidence and perseverance make them more likely to put themselves in challenging situations that create opportunities for flow. Mental toughness is also correlated with physical toughness and tolerance of pain. Experiments have found, for example, that mentally tough sports students perform better on a test requiring them to hold up a heavy weight for as long as possible.[11]

Resilience is not limitless, of course. Severe or prolonged stress will trigger a range of psychological and physiological responses in anyone, as we described in Chapter 1. These responses are normal and generally reversible. In some circumstances, however, severe or prolonged stress may have longer-term negative consequences, ranging from what is popularly known as 'burnout' to PTSD.

For some people, spending long periods of time in extreme environments, such as on polar missions or long-duration space missions, is associated with harm, usually temporary, to mental health. Instances of depression, concentration and memory problems, and irritability have been documented in astronauts who spend weeks or months in space.[12] In a study of winter-over crews in Antarctica, psychiatrists assessed that one in twenty had developed symptoms of serious psychiatric disorders during their stay.[13] More common, but less severe, is 'winter-over syndrome'. This is a cluster of symptoms, including irritability, social withdrawal, cognitive impairments, and depressed mood, accompanied by insomnia and a tendency to stare blankly into space.[14]

Winter-over syndrome has been described as a kind of burnout.[15] Burnout may be regarded as a breakdown of resilience in the face of work pressures. Its three main features are exhaustion, cynicism, and loss of performance. As with winter-over syndrome, the adverse changes that characterize burnout are destructive, although they fall far short of PTSD. Research suggests that individuals are more likely to suffer burnout if they feel they have no control over what happens to them in the workplace.[16] As we saw in Chapter 1, control is an important mediating factor in stress: other things being equal, stressors are more stressful if the individual has little or no control over them.

In extreme circumstances, adverse psychological reactions to stressors persist and the individual may develop PTSD. Someone suffering from PTSD may have recurring, intrusive, and frightening memories of their trauma, persisting for months or years after a traumatic event. They also tend to experience heightened arousal, making them irritable, hypersensitive to sights and sounds, and unable to sleep well.[17]

The great majority of people do not develop PTSD, however, even when they have been exposed to extreme stress. For instance, a study of almost 10,000 UK military personnel who had been deployed to combat zones in Iraq and Afghanistan between 2003 and 2009 found that only one person in twenty-five (4 per cent) developed PTSD as a result of their experiences.[18] Similarly, when researchers looked at how residents of Manhattan coped with the trauma of the 9/11 terrorist attacks in 2001, they found that only 7.5 per cent had symptoms characteristic of PTSD a month after the event, falling to fewer than 1 per cent after six months.[19] These and other comparable studies suggest that, in general, fewer than one in ten people exposed to extreme trauma will develop full-blown PTSD.

Human history is rich with examples of immense resilience, including the extraordinary individuals described in this book. It would be a mistake, however, to suppose that resilience is a quality displayed only by exceptional people in extreme circumstances. In fact, resilience is a normal and remarkably common human quality. Most of us suffer some trauma at least once in our life, such as the death of a loved one, having a serious accident, seeing or experiencing violence, or living through a natural disaster. Most of us cope reasonably well with that trauma: we do not suffer severe or long-lasting damage, and we do not usually need medical help.[20] The normality of human resilience is sometimes overlooked because of the tendency of psychological research and media reporting to concentrate on the minority of cases where resilience breaks down.

What makes some individuals more resilient than others? The ability to respond resiliently to adverse circumstances depends on a range of personal attributes, including past experience and genetic makeup. Moreover, people can be resilient in different ways: the same stressful situation may evoke distinctive but equally resilient responses from different individuals. Resilience is a complex phenomenon.

A growing body of evidence suggests that genetic differences between individuals affect their resilience, and hence their risk of responding pathologically to trauma. Studies of pairs of identical and non-identical twins have found that mental toughness has a relatively high heritability, with around half of the variation between individuals in mental toughness attributable to genetic differences.[21] A study of inner-city residents who had been physically or sexually abused as children found that certain variants of a gene called FKBP5 predisposed them to develop PTSD symptoms in adulthood, whereas other variants of the gene offered increased protection. The FKBP5 gene is involved in hormonal feedback loops in the brain that drive the stress response.[22]

Neurobiological research suggests that resilience depends partly on how well the 'reasoning' centres in the individual's brain, especially the prefrontal cortex, communicate with the brain regions responsible for mediating emotions such as fear. Brain scanning of PTSD sufferers has revealed that when they are reminded of their original trauma, the activity in their prefrontal cortex tends to be unusually low, whereas activity in the amygdala, a brain region that mediates fear, is unusually high. People who have experienced trauma without developing PTSD appear to have a more active prefrontal cortex.[23] It seems as if the reasoning part of their brain may be more active in moderating the emotional response to the reminder of the trauma.

Research has highlighted the importance of a hormone called neuropeptide Y (NPY) as a biological marker of resilience. NPY appears to damp down activity in parts of the brain that are involved in responding to stress, including the amygdala. NPY has the effect of reducing subjective anxiety and improving cognition in stressful conditions such as combat. Individuals with a low-expression variant of the gene that codes for NPY have higher levels of anxiety and bigger brain responses to potentially stressful stimuli, such as threatening facial expressions. NPY levels were found to be higher in special forces soldiers who had been selected and trained to be resilient.[24]

Biological factors such as these may have an influence on an individual's resilience, but they do not by themselves determine how well someone responds to stress. As with any psychological phenomenon, the development of resilience depends on a complex interaction between an individual's biology and their experience. As we shall see later, the evidence shows that we can learn to become more resilient.

One of the most significant environmental contributors to resilience is social support. Personal relationships with family, friends, and colleagues can provide a powerful buffer against stress and play a central role in making someone more (or less) resilient. Laboratory experiments have shown that simply holding another person's hand can significantly reduce an individual's stress response.[25] Other research shows that the level of social support enjoyed by war veterans has a strong effect on their likelihood of developing PTSD.[26] People who are socially isolated, and those who have been severely abused or neglected in the past, tend to be less resilient and more vulnerable to PTSD.

Finding meaning in life, even in adversity, is another way of strengthening resilience. Some people do this by means of spiritual or religious practices. Studies suggest that people who actively practise their religion tend to have better psychological and mental health compared to those who do not (with the exception of those who believe in a punitive, all-controlling deity). One likely mechanism for this effect is the greater social support that accompanies regular attendance at religious services.[27] Another possible mechanism may be through a form of mindfulness meditation, which is practised—often in the form of individual or collective prayer—in many religions. We saw in the previous chapter that meditation can improve the ability to control attention; as we shall see shortly, it may also help to cultivate resilience.

It is not necessary to be religious to believe that life has meaning. The psychiatrist Viktor Frankl, who survived the concentration camps of World War II, emphasized the importance of finding purpose and meaning as a way of enduring unimaginable hardship. In the camps, Frankl realized that in order to survive he had to imbue his existence with meaning. He became determined to survive so that, after the war, he could help others understand the atrocities that had gone on in the camps. He set himself concrete goals, which helped him rise above the suffering, and invented his 'meaning therapy', which became influential in the post-war period. Frankl's approach emphasizes the importance of the human spirit, but it is not grounded in religious belief or practice.[28]

Humour is another contributor to resilience. The ability to see the amusing or absurd side of adversity can help us roll with the blows. In common with other positive emotions, the right kind of humour can offset the impact of anxiety, fear, and other negative emotions. Humour can also act as a social lubricant, reinforcing personal relationships and attracting social support. Research has found that the appropriate use of humour in the workplace is correlated with better working relationships, greater job satisfaction, and improved productivity. The most creative and successful parts of an organization are often the noisiest ones.[29]

The use of humour is pleasant and has no obvious disadvantages, provided it is appropriate for the context. However, one person's light-hearted banter may be someone else's irritant, and humour can do harm if it descends into bullying. In extreme situations, self-deprecating humour may be less effective as a coping skill than humour that helps people to distance themselves from the stressful situation.[30] Incidentally, humour should not be confused with laughter, which is a social signal. Laughter often occurs in situations that are far from light-hearted or humorous, such as when the person laughing feels embarrassed, socially subordinate, or anxious.[31]

Humour has often been instrumental in enabling people in extreme environments to cope with stress and bond in the face of hardship. As well as engaging in verbal banter, people in extreme environments are often prolific pranksters.[32] One naval officer recalled how the crew of his nuclear submarine coped with boredom and fear by playing practical jokes, such as repeatedly abducting a junior officer, mummifying him in duct tape, and hiding him in an obscure compartment.[33] An astronaut on Mir described how he would 'sneak up' on a colleague and then 'hover patiently until he turned around. I knew that I had gotten him whenever he would gasp and flail his arms backward'.[34]

Members of the British Antarctic Survey have been particularly notorious for their practical jokes. It became a tradition for those who had over-wintered to sabotage the base as their replacements arrived to take over. On one occasion, the new arrivals at an Antarctic base discovered a fish lodged in the ventilation system, green food colouring in the water tanks, and cubes of strong-smelling cheese secreted in the bar.[35]

Humour was a crucial survival tool for Brian Keenan and John McCarthy, who were taken hostage in Beirut by Islamic Jihad terrorists in 1986. Keenan was eventually released four and a half years later, but McCarthy remained in captivity for another year. Both men spent periods in solitary confinement, but for much of their captivity they shared a small cell. They were usually blindfolded, often beaten, and were moved frequently. Keenan's account of their experience, An Evil Cradling, memorably illustrates how they used humour to cope with their lengthy ordeal and fend off the debilitating despair that afflicted many other captives.

From the moment the two men met, their banter was often bracing—the 'insatiable hurling of dog's abuse at one another', as Keenan put it.[36] Three American hostages with whom they shared one prison looked on in confusion, unsure whether the abusive sentiments were genuine. Even in the darkest moments, they often found themselves laughing. In one of their many squalid prisons, the guards prayed out loud for long periods, reducing Keenan and McCarthy to uncontrollable laughter: 'Each time these prayers started we would be caught up in a whirlwind of giggling. We were unable to speak for laughter, unable even to look at one another'. This disconcerted the humourless guards, who were unable to understand how anyone in such dire circumstances could laugh. Keenan described their humour as a 'shield' and a 'counterpart to depravity'.[37]

In addition to its psychological benefits, humour might also contribute to physical well-being (although the evidence for this is not conclusive and has in the past been somewhat overstated). There is empirical evidence that humour can increase people's tolerance of pain, and that such effects cannot be explained simply by distraction. Moreover, exposing people to humorous videos has been shown experimentally to evoke a short-term reduction in circulating levels of the stress hormones adrenaline and cortisol.[38] Such effects may, however, be the result of a broadly positive emotional state, rather than specifically humour.

The idea that positive emotions in general can improve resilience is well supported by research. For example, a US study of responses to the 9/11 terrorist attacks of 2001 found that people who experienced positive emotions, such as gratitude or affection, following the attacks responded more resiliently to the stressful experience. Positive emotions affect our cognition, mainly by broadening attention and ways of thinking, making us more flexible and creative. This can be helpful in a crisis. In contrast, negative emotions such as fear or anxiety have a narrowing effect on attention. Findings like these form the basis of psychologist Barbara Fredrickson's 'broaden-and-build' theory of positive emotions. Fredrickson has argued that actively cultivating positive emotions in the face of stress makes it easier to cope, both in the short term and longer term, by damping down arousal and making the experience more bearable.[39]

Not all of the qualities that make a person resilient under pressure are desirable in everyday life. One example is the personality trait of self-enhancement, which means the tendency to overestimate one's own abilities

and positive qualities in a self-serving way. This rather unattractive characteristic can be helpful in stressful situations. People with the trait are found to
cope better with stressors such as combat and bereavement.[40] In normal life,
however, they can appear narcissistic.

Cultivating resilience

How might someone go about strengthening their resilience? As we have
seen, there are many things a person can choose to do that are likely to
contribute to resilience, such as nurturing social relationships, finding meaning in adversity, and deploying humour.

Another approach to strengthening resilience is through the practice of
meditation. In Chapter 10 we outlined the basic nature of meditation and
explained how it improves the ability to control attention. The potential
benefits of meditation are more extensive, however, and they include
strengthening the ability to cope with anxiety, stress, and pain.

Evidence from observational and experimental studies shows that meditation can alleviate some of the psychological and physiological effects of
anxiety and stress.[41] For instance, a study of people who had been diagnosed
with anxiety disorders found that a programme of mindfulness meditation
produced significant improvements in their symptoms. Three years later,
most of them were still practising meditation and the clinical benefits had
been maintained.[42] Similar results have been found in numerous other
studies. A meta-analysis of twenty published studies showed that training
in mindfulness techniques produced significant benefits in terms of coping
with stress and alleviating symptoms in people with a wide variety of
problems, including chronic pain, anxiety, and stress.[43]

One study found that training in a mindfulness-based stress reduction
technique eased the symptoms of people with chronic anxiety and depression. In particular, it reduced their tendency to ruminate on problems.[44]
Rumination, in this context, means allowing one's attention to become
focused on a negative emotion like sadness, and worrying repetitively
about its causes and consequences. People who habitually ruminate in

response to negative moods often make those moods worse, and the tendency to ruminate is associated with vulnerability to depression. Mindfulness meditation helps to break this vicious cycle, as the practitioner learns to observe their thoughts and feelings from a more detached perspective. Other studies have found that people who meditate regularly tend to experience more positive emotions—which, as we saw earlier, are another contributor to resilience.[45]

Meditation can also mitigate some of the physiological responses to stress. In one experiment, participants practised meditation for five days, while control participants were given relaxation training. The meditators scored better on measures of anxiety, depression, anger, fatigue, and (as you would expect) the ability to control their attention. They also displayed a significant reduction in levels of the stress hormone cortisol.[46]

Extensive evidence shows that mindfulness meditation helps people to cope with both chronic and acute pain.[47] The benefits last for months or even years. For example, patients suffering from chronic pain found that their condition improved significantly during a ten-week programme of mindfulness meditation. By the end of the programme, two-thirds of the patients showed a reduction of at least a third in their average pain rating. These reductions in perceived pain were accompanied by large and relatively stable improvements in mood and psychiatric symptoms.[48] Another study of patients with chronic pain found that mindfulness meditation led to significant improvements in several outcome measures, including the extent to which pain inhibited everyday activity, anxiety, and depression.[49] The meditation patients in this study compared favourably to a control group who received traditional treatments for chronic pain, such as nerve blocks, physical therapy, analgesics, and antidepressants.

Acute pain can also be alleviated by meditation. In a series of experiments, volunteers were asked to rate their subjective experience of pain when given mildly painful electric shocks. Some participants then received three 20-minute lessons in mindfulness meditation. When further shocks were administered, those who used the meditation technique reported significantly reduced pain. In a related experiment, in which heat was used to cause localized pain, meditation substantially reduced the perceived pain.[50]

How does meditation help people cope with pain? As we saw in Chapter 10, the primary feature of mindfulness is the ability to observe one's thoughts, emotions, and physical sensations with a sense of neutral detachment. This mindfulness state appears to help people cope with pain by uncoupling the primary physical sensations from the emotional and cognitive reactions that normally ensue, thereby reducing the experience of suffering. Consistent with this theory, brain-scanning studies have shown that meditation reduces the level of pain-induced neural activity in the prefrontal cortex.[51]

Another way in which meditation may enhance resilience is by boosting aspects of immune function. For example, researchers found that healthy people who had practised mindfulness meditation for two months mounted a significantly bigger antibody response to an influenza vaccine.[52] Other research has revealed that practising meditation is associated with potentially beneficial alterations in the pattern of activity in genes, including genes associated with the control of blood sugar levels and inflammatory stress responses.[53]

In view of its considerable benefits, mindfulness meditation is widely taught to cancer patients, prisoners, and people suffering from anxiety, depression, or chronic pain.[54] However, the evidence implies that meditation can be of benefit to all of us, by helping to improve our ability to cope with stress. The added attractions are that it is safe and costs nothing.

Drugs

Despite the proven value of social support, humour, and meditation, some people habitually choose chemical remedies for their stress. Alcohol, nicotine, cannabis, and other psychoactive drugs can help to relieve acute anxiety and thereby bolster short-term resilience. And, of course, alcohol and opiate drugs have also been used for thousands of years to relieve pain. Laboratory experiments have shown that a couple of stiff drinks will reduce pain sensitivity by about the same amount as a dose of codeine.[55] The repeated use of alcohol and other psychoactive drugs carries risks, however, and often makes matters worse in the long run.

Even mild intoxication can of course be dangerous in extreme environments because of its effects on judgement and coordination. Even so, space agencies have debated the pros and cons of allowing alcohol during space missions, with some experts in the European Space Agency arguing that banning alcohol might deter people from volunteering for long-duration missions.[56]

Although outer space remains largely alcohol-free, the organizers of expeditions to most other extreme environments have long recognized the value of alcohol, nicotine, and other recreational drugs in relieving boredom and anxiety, and providing essential dollops of pleasure. Some also believed it had social benefits: 'Two men who have fallen out a little in the course of the week are reconciled at once by the scent of rum; the past is forgotten, and they start afresh in friendly co-operation', wrote Norwegian polar explorer Roald Amundsen.[57]

According to one source, the seventeenth-century explorer George Weymouth budgeted less for navigation equipment than he did for beer when planning his voyage to the North West Passage.[58] Centuries later, the organizers of the Ronne Antarctic Research Expedition were not generous enough for some team members, who reportedly made their own liquor by mixing ethyl alcohol, water, and lemon extract to produce a concoction they called 'sclaunch'. Meanwhile, the ship's doctor smuggled spirits on board by decanting them into medicine bottles that he hid in his office. One member of the expedition remarked that drinking was a safety valve, adding that 'merely knowing that the alcohol was available in a crisis drained off tensions, made frictions between individuals more bearable'.[59]

A poignant example of mood-altering drugs providing relief in the face of extreme stress comes from the aftermath of the 1986 nuclear disaster at Chernobyl. A volunteer team of Soviet scientists undertook a series of extremely dangerous missions into the concrete 'Sarcophagus' that covered the highly radioactive remains of the damaged reactor. The men had no proper protective equipment, no prospect of financial reward, and no illusions about the mortal risks they faced from radioactive fuel and dust. They repeatedly exposed themselves to levels of radiation that would be considered almost suicidal in the West, while protected only by cotton wool masks. A few of the original scientists could not cope with the stress and had to leave. To relieve the tension when they were about to enter the Sarcophagus, the remaining scientists drank alcohol and smoked. Several of them died from stress-related heart attacks or strokes before the radiation exposure had time to give them cancer.[60]

Toughening up

Stress is not all bad, despite its unremittingly negative image. In fact, some forms of stress can make us stronger, both physically and mentally. We all know that, at the physical level, our muscles and bones grow stronger as a result of being worked hard. Repeated bouts of physical exercise have a training effect, in which the body overcompensates by building stronger muscles and bones that are capable of coping with bigger demands in the future. But only up to a point, of course: apply too much stress and muscles will tear or bones will break.

Conversely, lack of use will make muscles and bones weaker. If you lie in bed for a month, your muscles will atrophy and your bones will become less dense. This phenomenon remains a challenge for the crews of long space missions, whose muscles and bones deteriorate as a result of prolonged under-use in microgravity.[61]

In analogous ways, coping successfully with psychological stress can make us mentally tougher and more resilient. Over a century ago, American experimental psychologists Robert Yerkes and John Dodson discovered an inverted U-shaped relationship between arousal (or moderate stress) and cognitive performance. They showed that people often perform best when exposed to moderate amounts of psychological stress. Very high and very low levels of arousal are associated with poorer performance.[62]

Since then, a large and growing body of evidence has shown that mild to moderate stress can have longer-term beneficial effects. In the 1950s and 1960s, experimental studies found that animals that were mildly stressed when young went on to have consistently lower levels of stress hormones, smaller and shorter physiological responses to stressors, and calmer behavioural responses to challenging situations. These effects can be long lasting.[63]

Repeated exposure to stressors can similarly make humans more resilient, especially when the stressors are acute, controllable, and moderate in intensity. This effect is known as stress inoculation, or toughening.[64] There is persuasive evidence that frequent biological arousal, similar to that of the short-term stress response, can improve an individual's mental and physical ability to cope with stressors, leading to improved performance under stress and greater emotional stability.[65] (Conversely, stressors that are chronic, uncontrollable, and severe will often damage the individual's capacity to cope with further stress, as well as harming mental and physical health.)

Moderate acute stress can also boost aspects of immune function, potentially making us more resistant to illness. It may even help to moderate the effects of aging—in contrast to chronic stress, which is associated with accelerated biological aging.[66]

A strengthening of resilience can occur even when the stress has been so severe as to cause some lasting psychological harm. A study of US veterans from World War II and the Korean War found that those who had survived the heaviest combat became psychologically more resilient as they grew older, compared with veterans who had experienced light combat or no combat at all. But they were also more likely to suffer emotional and behavioural problems, some of which were long lasting. When these men were followed up in their sixties, many described how their combat experiences had helped them become self-disciplined and cope with adversity. However, they also described negative effects, including persistent traumatic memories. Extreme experiences can have both positive and negative consequences, even within the same person.[67]

The writer Nassim Nicholas Taleb has coined the term 'antifragility' to encapsulate the broader concept that complex systems—including biological organisms such as humans, but also economic systems—become tougher as a consequence of coping with moderate amounts of stress. Taleb draws a sharp distinction between resilience, in which the person or system copes with stress but remains the same, and antifragility, in which the stress makes them stronger. As he puts it: 'Some things benefit from shocks; they thrive and grow when exposed to volatility, randomness, disorder, and stressors'.[68] Such systems are, in Taleb's terms, antifragile rather than merely resilient. They are like the Hydra of Greek mythology: each time one of its snake-like heads is cut off, two more grow back, making it even stronger.

An apparent example of a physical toughening effect, or antifragility if you prefer, derives from being intermittently deprived of food. Extreme environments often expose people to lack of food, whether through periodic acute starvation or prolonged shortage. Psychological and physical health are of course undermined by chronic malnutrition. However, intermittent acute reductions in food intake appear to have the opposite effect, in much the same way that moderate acute stress can be beneficial. In particular, there is growing evidence that intermittent fasting can make us healthier and physically more resilient.

As far back as the 1930s, scientists found that severely restricting the food intake of rats and mice made them live longer.[69] Scientists have also known for a century that significant food restriction can impede the growth of tumours in animals and reduce the incidence of cancer.[70] The reduction in calorie intake in such studies has generally been substantial and continuous, although not so severe as to cause malnutrition.

In the 1990s, an unintended human experiment lent further support to the idea that a restricted diet may be beneficial for health. Conceived by a group of eco-enthusiasts, Biosphere 2 was a two-year isolation study of eight men and women housed in a giant bubble. A sealed glass and steel structure enclosed 3 acres of artificial closed system in Arizona, designed to replicate natural ecological processes. Different sections of the habitat simulated a shallow ocean, desert, savannah, intensive agricultural land, and rainforest.[71] From the moment the door closed behind them in 1991 until the day they stepped out again two years later, the eight 'Biospherians' had to survive on what was contained within their habitat. They brought some supplies with them (including alcohol and chocolate), but once they were inside nothing was allowed in or out. Food had to be grown and there was no exchange of air with the outside world.

The physical conditions were challenging. The team spent much of their stay in a state of mild oxygen depletion, which affected their sleep and cognition. They were hungry and lost weight. Their food quickly grew boring. Hard physical work was needed to stay alive, and their daily intake of less than 2,000 calories was meagre. Nonetheless, they emerged in remarkably good health. As well as being slimmer and generally free from illness, the Biosphere 2 crew emerged with lower blood pressure, lower blood insulin and glucose, and lower cholesterol, compared to pre-experiment measurements.[72]

More recently, scientific and public attention has focused on the benefits of intermittent fasting, in which calorie intake is restricted on, say, two days a week. Someone following an intermittent fasting regime can eat as much as they like on 'feast' days, while substantially reducing (but not eliminating) their intake on fast days. Mounting evidence from studies of humans and other species suggests that this type of regime can bring significant health benefits. Some of these benefits result from weight loss and the associated improvements in risk factors for type 2 diabetes and cardiovascular disease.[73] But there seems to be more to it than losing weight.

Intermittent fasting and sustained calorie restriction both appear to trigger physiological defence mechanisms that make cells more resistant to biochemical stress. They also appear to boost levels of cellular growth factors that protect nerve cells against damage, thereby potentially reducing the risk of neurodegenerative diseases.[74] In some respects, the effects of calorie restriction or intermittent fasting are similar to those of physical exercise.[75] In both cases, repeated exposure to a moderate, acute stressor is sufficient to provoke a protective response, without causing lasting damage.

The apparent benefits of intermittent fasting are less surprising when viewed from an evolutionary perspective. Throughout most of our species' history, the natural environment in which humans evolved has been one in which food supplies were uncertain and calorie intake was liable to fluctuate markedly over time. Biological mechanisms consequently evolved to cope with such stressors. Most adult humans can survive for weeks without food and suffer no lasting damage. Alternating between plenty and hunger may be a more natural state, in evolutionary terms, than inhabiting an environment in which highly palatable food is always superabundant.[76]

Intermittent fasting may therefore be one example of a broader principle—namely, that repeated exposure to moderate, short-lived, and controllable stressors can make us tougher and more resilient, both physically and psychologically.

To summarize: stressful experiences can be traumatic, but they are also capable of conferring benefits. Although trauma can cause long-term damage to mental and physical health, this is not inevitable or even usual. Resilience is a common quality. Most of us encounter traumatic events during our lives and most of us cope reasonably well. The experience of repeatedly coping with mild to moderate stressors tends to make us tougher and psychologically more resilient. Other factors that bolster resilience include social support, humour, meditation, and the ability to find meaning in adversity. (Even drugs have their place.) Some stress is good for us.

12

Choosing Extremes

Some thousand miles up a river, with an infinitesimal prospect of returning!
I ask myself 'Why?' and the only echo is 'damned fool!...the Devil drives'.

Sir Richard Burton, British explorer (1821–1890)

What makes someone step voluntarily into harm's way, knowing they will encounter hardship, fear, and possibly death? In this chapter we explore the motivation of the people who choose to enter extreme environments. Why do they decide to go to such hard places and why do many of them keep returning? For those who have never experienced extremes, it can be difficult to imagine what drives these extraordinary individuals—people like Joe Kittinger, for example.

In the early morning of 16 August 1960, Kittinger did something that most of us would never contemplate. He stood alone in a cramped metal capsule suspended below a helium balloon. The blackness of space yawned above him and the desert of New Mexico lay 20 miles below. Muttering a prayer, Kittinger stepped over the edge and hurtled into space. He was in free fall for four and a half minutes travelling at more than 600 miles an hour before deploying his parachute. The descent took him almost 14 minutes. It was one of three high-altitude jumps that Kittinger made for the US Air Force to investigate whether humans could survive such conditions. He had volunteered—enthusiastically—for the work.

To attempt such a feat is to flirt with disaster. Equipment failure or a poor decision could have brought any of Kittinger's jumps to a tragic end. In fact, this was nearly the outcome of his first jump, when he almost died after choking on a tangled parachute cord. During his record-setting jump, one of Kittinger's pressure gloves malfunctioned, causing his right hand to swell agonizingly to twice its normal size and rendering it useless during his descent.[1]

A common belief is that people like Kittinger, and others we have mentioned in this book, are 'dare-devils' who risk their lives because they enjoy the thrills. Some individuals are indeed motivated by a desire for extreme sensations, as we shall see shortly. But there are many motives for choosing extremes, and each person has their own complex reasons. People choosing the same extreme activity do so for different reasons, while those choosing different activities might do so for similar reasons. One deep-sea diver might be motivated primarily by the pursuit of scientific knowledge, whereas another might dive mainly for the satisfaction of overcoming the physical challenges. A solo sailor and a skydiver might both be chasing the social status that comes from breaking records, and so on.

The differences in people's motivations for engaging in extreme activities reflect the differences in their individual life histories. The courses of our lives are shaped by genetic predispositions, personality traits, chance events, expectations, and above all our experiences, especially in childhood and young adulthood.

We are all influenced by expectations—our own expectations, built on our sense of personal identity, and the expectations of others, including families, peers, and wider society. These expectations are themselves influenced by the culture in which we are embedded. At every step of our lives we make choices, big and small, about what we start doing, what we continue doing, and what we stop doing. These choices are shaped by factors over which we have varying degrees of control, and which open up or close down different possible futures. Each of us has abilities, opportunities, and chance experiences that create options unavailable to others.

Our choices lead to outcomes, some of which are psychologically rewarding, others of which are aversive. Other things being equal, we tend to carry on with activities that are rewarding and avoid those that are unpleasant. But individuals differ in what they find rewarding, which means that two people who engage in the same activity can find the subjective experience to be very different. Moreover, people often endure unpleasant experiences in order to attain long-term goals, such as contributing to scientific knowledge or getting their name into the history books. An experience can be unpleasant yet ultimately satisfying.[2]

As a result of their experiences, individuals develop beliefs about themselves, and those beliefs shape their future choices. The multifaceted accumulation of predispositions, experiences, beliefs, and attitudes is, psychologically speaking, what makes you who you are.

It would take far more than a single chapter to do justice to the huge body of research and theories about why people behave the way they do. We will therefore focus here on a few themes that are particularly relevant to extreme environments. We start by considering the role of personality.

Personality

Each of us has a distinctive way of interacting with the world, which is commonly referred to as our personality. Broadly defined, personality can be understood as a set of characteristics that account for consistent patterns of feeling, thinking, and behaving.[3]

Many theories hypothesize that personality can be described in terms of a set of dispositions, or personality traits. Some psychologists argue that these traits explain our patterns of behaviour and thinking, while others maintain that traits are merely descriptions. In recent years, research has begun to uncover intriguing links between genetic and neurobiological differences and individual patterns of thoughts or behaviour. These links are helping to make sense of the biological basis of personality traits. However, disagreements remain over how many fundamental traits make up our personalities, and what those traits are.[4]

When considering extreme environments it is easy to make assumptions about personality, which on closer examination do not stand up to scrutiny. Take, for example, one of the best-researched personality dimensions: introversion–extraversion. Extraversion as a trait appears in all established psychological models of personality, and there is considerable evidence that it has a biological basis. The concepts of introversion and extraversion long ago escaped the confines of academic psychology and are widely used in everyday conversation, albeit in ways that do not always reflect the psychological definitions.

Broadly speaking, individuals who score highly on measures of extraversion tend to seek stimulation, whereas those who score low tend to avoid it. When asked to describe a typical extravert, most people tend to think of the lively 'party animal', equating extraversion with a preference for social interactions. However, individuals who score highly for extraversion seek more than just social stimulation: they also tend to gravitate towards other stimulating situations, including active leisure and work pursuits, travel, sex, and even celebrity. Introverts, on the other hand, have a generally lower affinity for stimulation. They find too much stimulation, of whatever type, draining rather than energizing. Contrary to popular belief, introverts are not necessarily shy or fearful about social situations, unless they also score highly on measures of social anxiety and neuroticism.[5]

On this basis, one might assume that extraverts would be drawn to extreme environments, where they could satisfy their desire for stimulating situations, whereas introverts would find them unattractive. And yet, as we have seen, extreme environments may also expose people to monotony and solitude—experiences that extraverts would find aversive, but which are tolerated or even enjoyed by well-balanced introverts. The point here is that simple assumptions about broad personality traits are unlikely to provide good explanations of why people engage in extreme activities.

Another well-researched personality trait that is relevant to extreme environments is sensation seeking. Marvin Zuckerman, the psychologist who developed the concept, has defined sensation seeking as 'the seeking of varied, novel, complex and intense sensations and experiences and the willingness to take physical, social, legal and financial risks for the sake of such experiences'.[6]

The trait is measured using a psychometric tool called the Sensation Seeking Scale (SSS). Each item on the questionnaire requires the respondent to choose between a cautious preference, such as 'I prefer quiet parties with good conversation', and one involving a desire for sensation, such as 'I like wild and uninhibited parties'. The SSS is made up of four subscales, each of which is intended to capture one particular dimension of the trait. They are: disinhibition (a desire to seek stimulation through partying, drinking, and sexual variety); thrill and adventure seeking (a desire to engage in risky and adventurous activities and sports that provide unusual sensations); experience seeking (the pursuit of new sensations through music, art, travel, or psychoactive drugs); and boredom susceptibility (an aversion to monotony).

Sensation seeking is sometimes confused with impulsiveness, but the two traits are distinct. Whereas sensation seeking refers to a preference for novel and intense experiences, impulsiveness is about the ability to control behaviour. A high level of sensation seeking might lead someone to take risks because of the thrill this provides, whereas impulsiveness leads them to take risks because they lack the self-control to stop themselves. An individual who scored highly on both sensation seeking and impulsiveness would be even more likely to take risks.

Research has revealed consistent associations between the trait of sensation seeking and a wide range of risky activities, including dangerous driving, illicit drug use, risky sexual behaviour, and criminality, especially among young adults. Individuals with high sensation seeking scores are empirically more likely to drive aggressively, engage in speeding, violate traffic regulations, drive while drunk, not wear a seatbelt, and (inevitably) have traffic accidents.[7] Other evidence suggests that engaging in reckless sensation seeking might also be a form of behavioural self-medication, which enables individuals to distract themselves from unpleasant feelings of depression or anxiety.[8] (We touched on this idea in Chapter 2, when we considered people who found that extreme activities such as diving and mountaineering helped them escape from the chronic anxieties of their everyday lives.)

Anecdotal evidence suggests that at least some of the people who choose extremes do behave in ways that are consistent with a sensation seeking personality. For instance, it is said of the great polar explorer Sir Ernest

Shackleton that he was a showman who 'drank too much, smoked too much and slept with other men's wives'.[9]

Sensation seeking was clearly visible in the behaviour of Commander Lionel 'Buster' Crabb, a leading military diver of his day.[10] Before World War II, Crabb had a number of exotic jobs, including male model, gunrunner, and spy. When war broke out he volunteered to be a navy bomb disposal diver, despite being a poor swimmer with bad eyesight who hated physical exercise. Crabb was notorious for ignoring safety precautions and being uninterested in how his diving equipment worked. Nonetheless, he became one of the Royal Navy's greatest combat divers and was awarded the George Medal for repeated acts of extreme bravery. He disappeared under mysterious circumstances in 1956, while reportedly carrying out a secret diving mission on behalf of British Intelligence to inspect a Soviet warship.

'Buster' Crabb was a quintessential sensation seeker, with appetites that included risk-taking adventures, sex, and drugs. He was a chain-smoking heavy drinker with a passion for gambling and women. His ex-wife claimed he had a sexual fetish for rubber, which might have been one of his more exotic motivations for diving. Crabb is said to have worn a pink rubber mackintosh underneath his navy uniform, which made him 'rustle like a Christmas tree'.[11]

As with extraversion, a link between the sensation seeking trait and a preference for extreme environments makes intuitive sense. Moreover, there is empirical evidence to support it. Studies have shown, for example, that people who engage in extreme sports such as hang gliding or surfing tend to score higher on sensation seeking than those who engage in less risky sports such as golf.[12] That said, the relationship between sensation seeking and extreme activities is not clear-cut.

According to the evidence, most people who operate in extreme environments are not big sensation seekers (and neither are they impulsive). To give one example, researchers who studied the personality characteristics of participants in a hazardous expedition to the North Pole found little evidence of sensation seeking tendencies. The expedition members did, however, display high levels of self-control and achievement orientation, along with low reactivity to stress.[13] A study of mountaineers found that, although they had higher overall scores on sensation seeking than a non-climbing control group, the differences were limited to two of the four dimensions of the trait—namely, thrill and adventure seeking, and experience seeking. The climbers did not differ from controls on disinhibition or boredom susceptibility.[14]

The research evidence that individuals who choose extreme environments are not simply sensation seekers is supported by anecdote and simple logic. Extreme operators such as divers and mountaineers depend for their survival on meticulous preparation, the avoidance of unnecessary risks, and the ability to remain unperturbed by acute peril. Those who do push the limits in search of thrills tend to die. We spoke to one climber who laughed when we mentioned the popular stereotype of climbers as thrill seekers. 'Control freaks, more like', he told us. 'We try to eliminate the thrills, not look for them'. He reminded us of the climber's adage that 'there are old climbers and bold climbers. But there are few old, bold climbers'.

At first sight, then, we have a paradox. A desire for sensation would predispose someone to seek out an extreme environment, but the evidence suggests that many of the people who do so are not in fact sensation seekers. One partial explanation is that extreme activities differ in the rewards they offer. People who are high in the need for sensation may be attracted to some sorts of extreme activities but not others.[15] Some activities, such as skydiving or BASE jumping, offer short bursts of intense sensation, whereas others, such as climbing or diving, offer sensations that are more prolonged but often less intense.

We must also distinguish between the psychological factors that shape someone's initial choice of activity and the factors that subsequently under-pin their sustained engagement in that activity. A sensation seeker might initially be attracted to an extreme environment, but their ability to survive and thrive there will depend on other qualities. In line with this distinction, research has found that individuals who score high on sensation seeking tend to engage in a wider range of activities than low sensation seekers, but for less time.[16] Those for whom sensation is the most attractive aspect may not stay for long, while those who remain and become veteran practitioners are probably motivated by something else.

Another, more prosaic, reason why many of the people who enter extreme environments are not sensation seekers is that they have been carefully selected from a large pool of applicants, as happens for polar or space missions. Selection processes are generally designed to weed out impulsive risk-takers with the 'wrong stuff', who might jeopardize the mission.

Development

Personality traits account for some predispositions towards extreme environments, but they are certainly not sufficient to predict who will actually end up there. The course of each person's life is shaped by their experiences and the circumstances they encounter along the way. Family history and childhood experiences, in particular, can kindle or nurture an interest in extreme environments.

Children, in common with young animals of other species, naturally engage in playful behaviour and exploration. This can be highly physical, including play fighting and rough-and-tumble play. Biologists believe that young animals, children included, acquire valuable skills and experience through play, as they engage in demanding activities while being relatively insulated from the normal consequences of their behaviour.[17] Play allows children to stretch themselves and learn how to cope with mental and physical challenges. Childhood play may also give some individuals a taste for adventure, which they carry into adult life.

Many people have childhood experiences that foreshadow an adult passion for adventure. Wilfred Thesiger's life of exploration started before he was 1 year old, when his mother journeyed with him across Ethiopia. It was the first of many long journeys that Thesiger made, usually on camels or mules, during his childhood. He retained vivid memories of his childhood in Africa and believed they had contributed to his enduring interest in travelling.[18] Salomon Andrée, who led the first (and doomed) attempt to reach the

North Pole by balloon in 1897, launched his first gas balloon when he was a boy. That voyage also ended in disaster when the balloon caught fire.[19]

Some adventurers appear to have been driven by a desire to live up to the expectations of their parents. As a child, Apsley Cherry-Garrard listened devotedly to stories of his high-achieving father's courage, and felt pressure to match his father's example. He was 21 when his father died, and three years later he applied to join Scott's expedition to the Antarctic.[20]

In the case of 1930s aviator Jean Batten, it was her mother Ellen who shaped her destiny. Jean's biographer describes Ellen as a highly controlling mother who focused obsessively on Jean, at the expense of her marriage and her relationship with her other children.[21] From the moment Jean was born, Ellen was convinced that her daughter would be an international superstar. With her mother's encouragement and help, Jean became a talented and supremely confident pilot who set several records for solo long-distance flight. Unfortunately, her mother's single-minded cultivation of Jean's self-esteem created an arrogant and obnoxious show-off. One psychologist described Jean as showing all the signs of an extreme narcissistic personality.[22]

Parents are not the only role models, of course. Some people are inspired to strive for great achievements by someone they admire, even if they do not know them personally. Each generation has its inspirational heroes. A recent review of the experiences of modern members of the Explorers Club found that the writings of the flamboyant adventurer (and expert self-publicist) Richard Halliburton had been inspirational for one generation, while Thor Heyerdahl, who sailed his raft *Kon-Tiki* across the Pacific in 1947, had inspired the next.[23]

For some individuals, it was not conformity with their upbringing that shaped their path to adventure, but rebellion against it. Aleister Crowley was raised in the strict puritanism of a Plymouth Brethren household. In common with many adventurers, Crowley lost his father at a young age, and from that point on he seems to have done everything imaginable (and more besides) to distance himself from his restrictive background. He devoted his life to hedonism, occultism, drug taking, sex, international travel, and a flagrant disregard for the rules of society. To his great delight, Crowley was reviled by the press as 'the wickedest man in the world' and happily referred to himself as The Beast. His guiding principle in life was 'Do what thou wilt'.[24] Almost overlooked in this sensational and scandalous life is the fact that Crowley was also an exceptional mountaineer, who made the first serious attempt to scale K2 in 1902, reaching 22,000 feet before retreating.[25]

Chance and opportunity

Many people have personality traits, childhood experiences, and inspirational role models that might have instilled in them an urge for adventure. Yet the vast majority do no more than become armchair adventurers, or indulge in a little recreational risk-taking, perhaps on adventure holidays or weekend climbing trips. What, then, are the other features that distinguish those individuals who dedicate themselves to extreme activities from the rest of us who do not?

The experiences that shape an individual's development continue beyond childhood, of course. These experiences arise through deliberate choices, chance circumstances, or a combination of the two. Opportunities to engage in extreme activities may occur through chance contacts with other people. Friends or relatives may help by providing practical support, as well as encouraging the desire for adventure.

Individuals also create their own opportunities. The scientists who won coveted places on space missions could do so only after they had studied hard in their scientific fields. Before Joe Kittinger made his stratospheric parachute jumps, he was an experienced pilot who had flown experimental jet fighters throughout the 1950s. He could not have reached that position without persistence and hard work. Having become a pilot, Kittinger found himself working for an organization with the resources to send a man 20 miles into the air. And because he had become an exceptionally good pilot, he was the man the US Air Force chose for the task.[26]

In a few cases it is a lack of alternative opportunities that steers a person towards extreme environments. One example is Lawrence 'Titus' Oates, who died with Captain Scott on their way back from the South Pole in 1912. Oates was dyslexic and failed to obtain any educational qualifications, which meant that most of the professional careers open to gentlemen of his privileged background were closed to him. Instead, he focused on excelling in outdoor activities and joined the army, where he served with distinction and was wounded in battle. Volunteering for the Scott expedition could be seen as a natural progression for someone with that experience.[27]

Chance events influence the course of everyone's life, and more so than might be supposed. Most theories about how people end up in their particular career or chosen activity tend to focus on conscious choice and planning, neglecting the role of serendipity. However, studies have shown that chance is often an important factor in shaping the direction of people's careers and wider lives.[28]

Chance played a role in the chain of events that led to Charles Darwin joining the Royal Navy ship HMS *Beagle*, an adventure into extreme environments that was to shape both the course of his life and the future of science. When Darwin heard that the *Beagle* would be embarking on a round-the-world voyage of scientific exploration, and that an unpaid opportunity was available for an expedition naturalist, he was eager to apply. He faced a problem, however. The expedition leader was a staunch believer in the then-popular notion that a person's character could be deduced from their facial features. He took considerable persuasion to accept that someone with a nose shaped like Darwin's could have sufficient determination and energy for the voyage. Darwin later mused on how the *Beagle* voyage, which he described as 'by far the most important event in my life', had depended on such a trivial matter as the shape of his nose.[29]

Saving lives, curiosity, and boredom

Some authors have attempted to develop a classification scheme, or taxonomy, of the motives for exploring extreme environments. One example was Wilfred Noyce, a mountaineer who was part of the 1953 Everest expedition. A few years later he wrote a book in which he attempted to distinguish between different motives. His taxonomy included fame and fortune, displaying courage, the joy of physical expression, knowledge, 'pure' companionship, penance (the 'hair shirt'), and 'escape'.[30] We have already touched on some of these motives, including demonstrating courage and relishing hardship. Another motive is saving lives.

Humanitarian motives have driven people to expose themselves to immense risks and hardships in war zones and disaster areas. One example is the British surgeon David Nott, who has been travelling to dangerous places since the early 1990s to provide life-saving emergency surgery to victims of wars and natural disasters. Each year, he takes several weeks of unpaid leave from his surgical job in London to carry out his humanitarian work. Among many other extreme environments, Nott has worked in front-line conflict zones in Bosnia, Sierra Leone, Sudan, Chad, Iraq, Afghanistan, and Syria.[31]

Humanitarian aid workers are exposed to many extreme stressors, both acute and chronic. They risk being caught in cross-fire, and sometimes, as in the Syrian civil war, they are specifically targeted. This makes aid work exceptionally risky, as well as profoundly stressful. Unsurprisingly, aid workers are more vulnerable to mental health problems such as anxiety and depression. As in other arenas, the evidence shows that the impact of these stressors is offset to some extent by protective factors such as social support.[32]

Two other important motives for entering extreme environments deserve some consideration. They are curiosity and the escape from boredom. Many adventurers have been driven to seek out extreme environments because of a desire to acquire knowledge or because they could not bear the prospect of being bored.

When Charles Darwin joined the *Beagle* expedition he was not attracted by the hardships he would endure on the five-year voyage around the world, which included chronic seasickness, climbing the Andes, and travelling for weeks on horseback through wild country. Instead, he was driven by scientific curiosity about undiscovered flora and fauna, and what his discoveries could tell him about the origins of our world.[33]

Captain Scott's final Antarctic expedition similarly had science as its core mission, despite the subsequent misleading impression that it was more about patriotic endeavour. Few men have suffered more in the name of science than the three members of Scott's team who walked for five weeks in icy darkness to collect penguin eggs (as we described in Chapter 1).

Curiosity has propelled many other scientists into extreme places. As a child, Penelope Boston dreamed of going to Mars. She became a microbiologist specializing in the study of extremophiles: microorganisms that survive and thrive in extreme conditions. Many such organisms live in deep, remote caves filled with noxious gases and poisonous slime. So this is where Boston has conducted her research, in dangerous subterranean spaces, wearing full protective gear. She was terrified the first time she went into a cave, but learned how to cave safely so that she could continue her scientific work. In one sense, Boston is closer to understanding the experience of a mission to Mars than most scientists will ever get.[34]

Geologist Natalie Cabrol is another scientist whose work in Earth's extremes has added to our understanding of other planets. High-altitude volcanic lakes are believed to be a good analogue for historic conditions on Mars. Cabrol's studies have cast light on the life forms that survive in such hostile conditions, and how they do it. Her research demanded exceptional diving and mountaineering skills, including holding a world record for high-altitude free-diving. She described exploring a lake at 20,000 feet in the Andes, during which she endured extreme cold and hypoxia, as 'the most rewarding experience of my life so far'.[35]

Another scientist who repeatedly put himself in extreme situations in the quest for scientific knowledge was the biologist J.B.S. Haldane. During World War II, Haldane led research into how submariners could escape from crippled submarines. In the course of various experiments, Haldane subjected himself to extremes of pressure and cold, resulting in exploding teeth, burst eardrums, temporary paralysis, a violent muscular contraction that resulted in a dislocated hip and crushed vertebrae, and nosebleeds so frequent that a colleague claimed he could track Haldane around the laboratory by following the trail of bloody handkerchiefs. Haldane explained that he preferred to conduct such experiments on himself because a rabbit would not say how it felt, and experimenting on a dog would require 'a licence signed in triplicate by two archbishops'. He left his body for medical research, noting that it had already been used for that purpose during his lifetime.[36]

While curiosity pulls some people towards dangerous places, others turn to extreme environments in order to escape from boredom. As we saw in Chapter 10, existential boredom is an unpleasant emotional state. The desire to avoid or escape from it can be strong. One philosopher wrote that a person who cannot tolerate a certain degree of boredom 'will live a miserable life, because life will be lived as a continuous flight from boredom'.[37] Some people are easily bored and find it hard to tolerate. They are often highly motivated to do something—anything—that provides stimulation and gives their life a sense of meaning or purpose.[38]

Individuals who are easily bored are more inclined to seek out excitement, variety, and novelty. They are also prone to act impulsively, reflecting a reduced ability to inhibit responses.[39] These propensities are reflected in a higher incidence of risk-taking behaviour, as we saw in Chapter 10: boredom-prone people are more likely to drive dangerously, take drugs, gamble, and commit crimes, among other things.[40] The philosopher Bertrand Russell was not joking when he argued that half the sins of mankind are caused by the fear of boredom. Russell wrote: 'experience shows that escape from boredom is one of the really powerful desires of almost all human beings'.[41]

History provides many examples of individuals who were driven to excess by their loathing of boredom. One was the occultist and mountaineer Aleister Crowley, whom we mentioned earlier in this chapter. Crowley declared that it was boredom, or the dread of it, that drove him to take heroin.[42]

Another man for whom boredom was a spur was World War II triple agent Eddie Chapman.[43] Originally a professional criminal, Chapman volunteered his services to German intelligence early in the war. He was parachuted back into England in 1942 as a German spy, whereupon he promptly volunteered to the British. Chapman was recruited by MI5 who sent him back to Nazi Germany, where he pretended to work for the Germans but under secret British control. One of his many unique achievements was to be awarded the Iron Cross (Nazi Germany's supreme bravery award) for his supposed services to the Fatherland.

Chapman's handlers, both British and German, often remarked on his susceptibility to boredom, and how this spurred him into impulsive risk-taking. One of his MI5 minders wrote that he was 'a man to whom the presence of danger is essential'.[44] Before the war, Chapman had worked as a barman, film extra, masseur, dancer, boxer, and wrestler, before turning to crime. He acquired a taste for cognac, women, and gambling at an early age. Many women (and quite a few men) found him irresistible. After the war, Chapman returned to the London underworld and became involved in smuggling gold and cigarettes across the Mediterranean. He never settled into a conventional life.

A complex path

As we have seen, individual motivations are complex and inherently difficult to explain. The reasons why someone engages in a particular activity will be diverse. Moreover, the individual concerned may not be consciously aware of those reasons. None of us can be confident of really understanding our own motivation. A large array of factors, including personality, upbringing, experiences, opportunities, and a dash of chance, combine in many different ways to shape the course of each person's life. We conclude this chapter by looking at one particular example of how these factors came together for one person who died in the pursuit of extreme experiences.

Nick Piantanida had always pushed himself to his limits.[45] As a child in the 1930s and 1940s he had shown an early determination to try new experiences. Having chosen a new activity, he worked tirelessly to master it, whether it was karate, basketball, boxing, or climbing. While still a schoolboy he taught himself SCUBA diving, then a much rarer and riskier activity than today.

In his early twenties Piantanida focused on becoming the first person to scale the supposedly unclimbable north face of Venezuela's Devil's Mountain, over which cascades the world's highest waterfall, Angel Falls. Against all the odds, enduring weeks of extreme temperatures, continuous rain, and punishing jungle hardships, the self-taught climber succeeded in setting his record. He did not become famous or rich, but that did not matter to him. He was driven by the thrill of becoming the first to accomplish what others had seen as an impossible feat.

When he returned from his Venezuelan adventure, Piantanida did not choose a job that might have provided him with opportunities to develop into a professional adventurer. He had already served his time in the military and, besides, he had a track record from school onwards of avoiding 'official channels'. Instead, he took various conventional jobs, including truck driver, factory worker, and pet trader, pursuing risky adventures in his own time.

Piantanida's love affair with parachuting started by chance when he and his wife happened to drive past an airfield at the moment when a group of skydivers were dropping to earth. 'I need to learn to do that', he exclaimed. Parachuting was extremely dangerous in the 1960s: broken bones and fatalities were common hazards. Piantanida loved the challenge but wanted to push the sport to its limits. He therefore turned his attention to beating Joe Kittinger's altitude record. Following what was by now an established pattern of behaviour, Piantanida set out to learn all he could about falling from stratospherically high altitudes. In the words of his biographer, Piantanida transformed himself 'into the director of a one-man aeronautical research program'.[46]

Piantanida made three attempts at a stratospheric jump from a gondola carried beneath a helium-filled balloon. On his second attempt he reached a record-breaking height of 123,500 feet but was unable to jump because his oxygen hose stuck. On his third and final attempt, in May 1966, the 34-year-old accidentally depressurized his helmet at 57,000 feet. The gas bubbles that formed in his body following the sudden pressure drop caused irreversible brain damage. Although Piantanida was still alive when the gondola touched down, he never regained consciousness and died four months later.[47]

Why did Piantanida risk his life? There are many answers. One explanation is grounded in personality: Piantanida was by nature a thrill-seeker. He felt compelled to take risks: he would not, or could not, do anything in moderation. One friend remembered: 'Full head of steam, pedal to the metal. That was Nick. If you went with Nick to have a beer, you'd have six beers. Life to the fullest'.[48]

Another relevant aspect of Piantanida's personality was self-efficacy. He was supremely self-confident, which is understandable given that he had hitherto always triumphed over the odds. His previous experience reinforced his continuing risk-taking behaviour: whatever he had tried, he had succeeded in doing. His attention to know-how was also crucial. Piantanida knew what it took to be an expert in a risky endeavour and he displayed the persistence and focus necessary to master each activity he chose.

In terms of opportunities, Piantanida (unlike Kittinger) had to rely on his own wits to raise funds, recruit help, and develop expertise. He was clearly an exceptionally driven, intelligent, and capable individual—so, in one sense, the answer to the question 'why?' is 'because he could'. A risk-taker with less competence and determination could not have achieved what Piantanida achieved.

Another explanation for why Piantanida engaged in high-altitude jumps is that he was following a life-long pattern. When you have no reason to believe otherwise, your most reliable guide to what someone will do in the future is what they have done in the past. Piantanida's life had a consistent theme: that of an intelligent and persistent man who was full of curiosity and could not bear to sit on the side-lines, but had to experience things for himself. As he explained: 'Most people talk about such things and do nothing. I just have to go and see'.[49]

We have seen, then, that people put themselves in extreme situations for different combinations of reasons. An individual's choices are shaped by their personality, life experiences, and opportunities. Chance often plays a surprisingly important role. An aversion to boredom drives some people towards extreme situations, while others are attracted out of curiosity or a desire to save lives.

13

Staying and Leaving

I have more care to stay than will to go.
William Shakespeare, *Romeo and Juliet* (1595)

The extraordinary people whose exploits we have described are remarkable for choosing to enter extreme environments. They are perhaps even more remarkable for choosing stay and to keep on returning, exposing themselves to hardships again and again.

The reasons why someone starts engaging in an extreme activity are often different from the reasons why they subsequently persist with that activity. In this chapter we explore how people sustain and develop their commitment to extreme environments, why they eventually leave, and what happens to them when they do.

A concept that is central to understanding why people stay, and why they leave, is that of social identity—the sense of self, purpose, and belonging that we develop through being part of particular groups and holding particular roles.[1] Everyone who enters the world of extreme activities, including those who operate alone, develops a social identity as a member of an exclusive group. They come to define themselves by what they do ('mountaineer', 'cave diver', 'explorer') and how they do it ('expert', 'daring', 'record-breaking'). A strong sense of social identity is found in all demanding professions, including the armed forces, emergency services, medicine, and sports.

Each of us has multiple social identities—as a worker, parent, friend, and so on—and the same applies to the individuals who engage in extreme activities. When someone is so enthusiastic about one activity that it becomes central to their sense of identity, they can be said to have a 'passion' for that activity. Most of us balance our passions against our other roles and activities. For some individuals, however, their sense of self becomes so enmeshed with their passion for an extreme activity that other social identities are neglected or fail to develop.

Passion

Anyone who lacks passion is unlikely to hang around for long in hard places. Passionate involvement in extreme activities can bring many psychological rewards, including self-esteem, a sense of meaning and purpose, the experience of flow, and close friendships. But there is also a dark side. In some individuals, passion grows into obsession, leading them to take excessive risks and neglect their family and friends.

George Mallory had a deep passion for mountaineering and for Everest in particular. He tried three times to climb the mountain and died on his third attempt in 1924. Mallory's passion for Everest would eventually prove stronger than any concerns for his own safety or for the feelings of the wife and family he left behind. 'I can't tell you how it possesses me',[2] he once wrote to his wife. Mallory was unable, or unwilling, to explain his overwhelming drive to climb Everest. When a journalist asked why he was going back, he famously replied: 'Because it's there'.[3]

Psychologist Robert Vallerand, a leading expert on the topic, has defined passion as 'a strong inclination toward a self-defining activity that one loves, finds important, and in which one invests a significant amount of time and energy'.[4] Until Vallerand and his colleagues started exploring the nature of passion, psychologists had largely neglected it.[5] This is somewhat surprising, because most of us have at least one passion (as defined above). Vallerand's first study found that 84 per cent of a sample of 500 students claimed to have at least one passion, on which they spent an average of 8.5 hours a week and which they had been pursuing for an average of six years.[6]

According to Vallerand's theory of passion, the experience of repeatedly performing a rewarding activity leads to the formation of a psychological bond between the individual and the activity. Over time, the activity becomes internalized as part of the individual's identity. For most of us, this identification with a favourite activity is balanced against our other social identities. Vallerand calls this harmonious passion (HP). An individual with HP for a particular activity integrates that passion with the rest of their life. In some cases, however, a person develops obsessive passion (OP): an all-consuming focus on one activity and a self-identity that is dominated by that activity. Individuals with HP and those with OP often behave in similar ways, such as being persistent in their pursuit of their passion, but the psychological outcomes can be very different.[7]

To illustrate the differences between OP and HP, imagine two women, Olivia and Helen, both of whom are passionate about climbing. Olivia's passion is obsessive, whereas Helen's is harmonious. Olivia continues to climb under pressure from her father, who saw she was good at it and encouraged her to climb competitively. An obsessive passion like Olivia's could also result from peer pressure or from self-imposed pressure to boost self-esteem. Helen climbs because she finds it intrinsically enjoyable.

Helen and Olivia both value their own high performance and strive to become better climbers. Both women find climbing rewarding, but for different reasons. Helen's reward is the intrinsic pleasure of climbing, a sense of mastery, and the social benefits of spending time with like-minded people. Olivia is driven to perform better because she does not want to disappoint herself or her pushy father. Her OP is fuelled by a desire to beat others, avoid failure, and boost her self-esteem.[8]

People like Olivia and Helen often behave in outwardly similar ways. Psychologically, however, they are different, particularly when it comes to dealing with failure. In a series of studies, Vallerand and colleagues found that people with OP tended to improve their performance when prompted to think about the possibility of failure, whereas the performance of those with HP was unaffected.[9] People like Helen get pleasure from their chosen activity, but their sense of self does not depend on it. One bad day's climbing is of no great consequence. However, Olivia's self-identity is almost exclusively as a climber. So when she thinks about, or actually experiences, failure, it seems like a threat to her identity. This drives her to work even harder and achieve higher performance.

Both types of passion can motivate individuals to practise hard, and both can lead to spectacular outcomes. As Vallerand puts it: 'there are two roads to excellence'.[10] Studies have found both harmonious and obsessive passion among elite performers in a variety of fields including sport, music, and dance.[11] The primary goal of individuals with high HP tends to be the achievement of greater competence for its own sake. Those with high OP typically strive to be better than (or avoid being worse than) other people.[12]

The differences between HP and OP can affect psychological well-being. This has implications for everyone who has one or more passions—which is most of us. Vallerand found that people with OP tend to have poorer mental health and a higher incidence of depression and anxiety compared to those with HP.[13] The reason is probably related to the way in which OP can overshadow the rest of someone's life, especially their social relationships.

As we have noted in previous chapters, social support is a strong buffer against stress. Anything that undermines social support can therefore make an individual more vulnerable. OP is characterized by rigid persistence and compulsive engagement in the chosen activity: Olivia cannot bear to refuse an invitation to go climbing, so she ends up neglecting other activities and social relationships. Unsurprisingly, this has negative consequences, both psychologically and practically. Olivia feels guilty about neglecting other aspects of her life, but she is unhappy if she turns down an opportunity to climb. She cannot win. People with OP are more likely than those with HP to find that their passion erodes their relationships.[14]

The differences between HP and OP have implications for physical health too. When Vallerand's group studied dancers, they found that high levels of passion for dancing were associated overall with fewer injuries, because passion was linked to high levels of practice, fitness, and expertise. However, when dancers with HP and OP did suffer injuries, those with OP were much more likely to suffer continuing problems. This was because dancers with OP tended to soldier on, concealing their injury from others and thereby aggravating it, whereas those with HP tended to rest and recuperate after injury.[15]

People with high levels of OP are also more inclined to take risks. For instance, a study of cyclists found that those with OP were much more likely to cycle during dangerous winter weather compared to those with HP. People with HP can cope with abstaining from their passion because they have other things in their life, but OPs keep going even if it endangers their health.[16] OP is not all bad, however: people with OP perform well in highly competitive settings where performing better than others is critical to success.[17]

Passion is not the only reason why a person might persist with an activity, even when it is damaging their relationships and health. We all have a preference to behave in ways that are consistent with our previous actions or stated intentions.[18] If we say we are going to do something, we may find it difficult to back out without losing face or disappointing people who are important to us. In some extreme activities, this type of public commitment can become a dangerous trap. By his third and final parachute jump, Nick Piantanida was risking his life at least in part because he did not want to disappoint the sponsors he had worked so hard to recruit.[19]

Others have been trapped by hubris or self-delusion. In the 1930s the eccentric and pompous self-publicist Maurice Wilson announced that he would climb Everest to demonstrate the power of faith and fasting. He planned to crash an aeroplane into Everest and walk to the summit. The mission was doomed to fail, and Wilson probably knew this by the time he set out. He had no experience of climbing or flying and was hopelessly unprepared. Nevertheless, having generated so much publicity, he felt he had no alternative but to continue. Remarkably, he did eventually reach the slopes of Everest, but died there in 1934.[20]

Leaving extremes

Everyone who engages in an extreme activity must eventually stop. The way in which people cease their engagement lies on a spectrum. They may happily choose to withdraw, they may be encouraged or cajoled to withdraw, or they may have withdrawal forced upon them. The reasons for leaving are many and varied, and include a loss of opportunity, decline in physical or mental capability, disillusionment, burnout, injury, death, or simply deciding that there are more attractive alternatives.

Leaving an extreme activity, perhaps after many years of intense involvement, can mark a major change in a person's life. Some find this transition is smooth and adjust easily to a more conventional life. For others, however, the shift is difficult. Their social world and sense of identity are so bound up with their activity that they find it hard to adapt.

There is relatively little research on the transition away from extreme activities such as mountaineering, diving, or exploration. However, psychologists have examined transitions in other domains, including the armed forces, sport, crime, and conventional work, and their findings are relevant. These studies have highlighted several factors that affect how easily an individual is likely to adjust after leaving a role that has formed an important part of their life.

One of the most important factors is the degree of personal commitment to the activity. People tend to feel strongly committed to activities in which they have invested a lot of time and other resources, and in which they have achieved expertise. The more they have invested in an activity, the harder it is to walk away from. The associated social ties may also be hard to sever.[21] Another factor is the extent to which the activity has been a dominant part of the individual's life. A strong social identity that excludes other identities (as is the case with OP) can be problematic when the individual has to abandon the defining activity.

Sociologist Helen Ebaugh conducted groundbreaking research into how people in high-commitment roles experienced disengagement from those roles. Based on in-depth interviews with nearly two hundred 'exes', including an ex-astronaut, she found that the process of voluntary disengagement featured a number of turning points. Most significantly, Ebaugh's research showed that the ease of the transition depends on how successful the individual is in creating a new post-exit identity.[22]

The story of Buzz Aldrin, the second person to set foot on the Moon in 1969, shows how bumpy the transition can be.[23] Travelling to the Moon by means of 1960s technology was one of the most extreme of all extreme experiences, and everything in Aldrin's life had been focused on achieving that goal. Soon after returning, he wondered: 'What's a person to do when his or her greatest dreams and challenges have been achieved?'[24] At the age of 39, he had decades in front of him but no idea what to do with them.

Aldrin was highly committed to his role as an astronaut and had invested deeply in it. The rigorous selection and training had made NASA the centre of his life, with family taking second place. His exploits had been at the centre of the world's media and everyone regarded him as 'the astronaut Buzz Aldrin'. The sheer strength of his personal, social, and public identity as an astronaut made it difficult to move on to something new.

Aldrin became increasingly disillusioned with NASA, which he felt regarded him as little more than a public relations representative. He worked as a supervising flight instructor, appeared in TV commercials, and supported a range of charitable endeavours. Behind the public face, he struggled with depression and a sense of purposelessness. His domestic life was a mess. He sought psychiatric treatment, which helped in the short term, but over the following decades he battled with recurring depression and alcoholism.

When Buzz Aldrin eventually left NASA he did not have a conventional life to which he could return. His whole career had been in government service, first as an engineer and fighter pilot and then as an astronaut. It took him decades to find a stable new role. As an ambassador for space exploration, reigniting and redirecting his passion for space travel, Aldrin finally adjusted to life after the Moon.

Leaving a life-defining activity is likely to be easier if you can move on to an attractive alternative that meets your psychological needs. Joe Kittinger coped by becoming a civilian balloon pilot and later a mentor to the next generation of high-altitude skydivers—most famously Felix Baumgartner, who in 2012 broke Kittinger's record by jumping from a height of 24 miles. The Apollo 12 astronaut Alan Bean became an artist, although most of his paintings are of moonscapes and astronauts.[25]

Social support is another important factor. People with good social support are better placed to handle a difficult transition. Buzz Aldrin credited his third wife with providing the support he needed to overcome his mental health problems and settle into a post-NASA role.

Some individuals face involuntary exit from their defining activity because their employer has decided to drop them. Astronauts must be selected for each and every mission they make. At some point they will fail to make the grade, or must stand aside to make room for others. It can be hard to leave what one astronaut described as 'the coolest job ever'.[26] Similar selection processes apply to other extreme activities, including polar missions, commercial diving, and competitive sailing.

Involuntary career transitions are a feature of everyday life as well, most commonly through redundancy or illness. Long periods of unemployment are associated with poorer mental and physical health, with persuasive evidence of a causal link.[27] Even those who remain in good health can still suffer psychologically as a consequence of unemployment. Moving suddenly and involuntarily from a job that has been a defining feature of their life can easily undermine an individual's sense of identity and purpose.[28]

Injury can trigger an abrupt and involuntary transition. The risky nature of extreme activities means that individuals are sometimes forced to end their involvement suddenly and unexpectedly following an injury. Some of them recover from injury or return to the fray anyway. Climber Paul Pritchard narrowly escaped death when a falling rock smashed his skull during a climb in Tasmania. It took him months to speak again, and over a year to relearn how to walk. He never fully recovered from his head injury and remained partially paralysed, struggling with memory and speech defects. Nevertheless, he continued to put himself in harm's way, including crossing the Himalayas on a specially modified tricycle.[29]

A sudden departure may be prompted by something bad happening to another person. In one case, a parachutist gave up skydiving after the death of a colleague in a skydiving accident. She was particularly unnerved by the manner of her friend's death: 'I had prepared myself for someone to die, but the way I thought they were going to die was different to the way he died.

This was an extraordinary accident.... It's freaky'. Because she perceived that the freak accident could have happened to anyone, her sense of control was shaken and she could no longer participate.[30]

Withdrawal from an extreme activity may be gradual, in line with gently declining physical capabilities or shifting priorities. Or it may be spasmodic, with repeated departures and 'relapses', as was the case with climber Jim Wickwire. While climbing Alaska's Mount Denali in 1981, he and his partner Chris Kerrebrock fell into a crevasse. Wickwire managed to struggle out, even though he was severely injured, but Kerrebrock was trapped by his rucksack, jammed between the walls of the crevasse. After many hours of desperately trying to free his friend, Wickwire eventually left him to die. The grief-stricken Wickwire promised himself and his family that he would never climb again. In the decades that followed, however, he repeatedly broke and made the same promise, witnessing several more friends die and coming close to death himself. He likened his obsession with climbing to an addiction.[31]

A game of consequences

As we have seen, extreme activities have a range of potential consequences, both positive and negative. At worst, they can be fatal. When an adventurer dies, their loved ones are left behind to suffer. The climber Robert Macfarlane wrote of 'all those ruined lives which have to be completed' after someone dies in pursuit of extreme experiences.[32]

When NASA planned the 1969 Moon landing, they knew there was a significant risk that Neil Armstrong and Buzz Aldrin might be stranded on the Moon. White House staff prepared for that eventuality. Alongside a solemn speech, which President Nixon would have broadcast to a grieving nation, were some grim instructions: 'Prior to the President's statement, the President should telephone each of the widows-to-be'.[33]

The emotional impact of an adventurer's death can be even worse if their remains are never found, as sometimes happens in extreme environments. Anthropologists have argued that a distinction between a 'good death' and a 'bad death' is culturally universal. Broadly speaking, 'good deaths' are dignified, painless, prepared for, and humane. 'Bad deaths', such as suicides, murders, and accidents, are unexpected, unpredictable, untimely, or unjust. They leave the bereaved with a sense of something misaligned or unfinished. The worst 'bad deaths' are those in which the body is missing. All cultures have funeral rituals, which serve to underline the finality of death. Without a body, it is harder for the bereaved to start their transition towards acceptance.[34]

When Salomon Andrée flew a balloon into the Arctic in 1897, in an audacious but disastrous attempt to find the North Pole, one of his two companions was Nils Strindberg (cousin of the author August Strindberg). Nils was betrothed to one Anna Charlier, to whom he wrote increasingly tragic letters as the three men battled through awful conditions. Nils Strindberg was acutely aware of the effect his prolonged absence would have on his sweetheart: 'Poor little Anna, in what despair you will be if we should not come home next autumn. And you can imagine how I am tortured by the thought of it'.[35]

Strindberg and the others did not come home. Their bodies and the letters to Anna lay undiscovered for three decades. Anna never got over Strindberg's disappearance and for many years clung to the hope that he might return. When she died in 1949, the ashes of her heart were buried next to Strindberg's cremated remains in Stockholm.[36]

Psychologist Pauline Boss, who is an expert on 'ambiguous loss', has conducted research with families of people who are missing and presumed dead. Individuals who suffer this ambiguous loss are prone to depression and anxiety. They can feel helpless and confused, and find the unending uncertainty draining. Boss's advice to them is to recognize that such feelings are normal and to find ways to live with the ambiguity.[37]

Most people who have suffered ambiguous loss do eventually accept that their loved one has gone. Some find it helpful to use rituals as a way of symbolically letting go. After Pete Boardman and Joe Tasker died on Everest, their partners built a memorial cairn at the foot of the mountain, one of them describing it as 'the nearest thing we had to a grave'. Several years later, when Boardman's corpse was found, his widow burned a photograph of the body. This was, for her, the moment when he could finally 'disappear' from her life.[38]

Losing a family member can be particularly devastating for children, who may react with anger or denial. The daughter of American climber John Harlin fantasized that his death on the Eiger had been an elaborate fake and he would return to her. Mountaineer Alex Lowe's three-year-old son repeatedly buried toy people in a sand pit and dug them up again. Lowe had died in an avalanche.[39]

It would be misleading, though, to dwell only on death and suffering. For those who take part in extreme activities and live to tell the tale, the long-term psychological effects are generally overwhelmingly positive. Psychologists have found that individuals who survive extreme experiences generally emerge relatively unscathed and look back on their experiences with satisfaction. They tend to be more tolerant of stress, and many of them feel a deep sense of satisfaction.[40] Moreover, a growing body of research shows that their experiences at the time are predominantly positive as well.

Historical and contemporary accounts of polar exploration are replete with comments about feelings of serenity and personal growth.[41] For example, Danish military officers who spent months in isolation patrolling the Arctic wastes reported feeling a greater appreciation of life and of their own personal strengths.[42] A study of people who overwintered in Antarctica found that as the environment became harsher, the people living there experienced *less* depression, anxiety, loneliness, or lack of concentration.[43]

Astronauts' accounts contain three times as many references to positive emotions as to negative ones. From the earliest days of space travel, astronauts have been awe-struck in particular by the beauty of Earth as seen from space.[44] For some this triggers a profound reappraisal of their life: Apollo astronaut Edgar Mitchell, for instance, described his return journey from the Moon as a 'grand epiphany accompanied by exhilaration'.[45]

Extreme sports are similarly associated with positive psychological outcomes.[46] BASE jumpers have described themselves as becoming more courageous and more humble as a result of their experiences. One BASE jumper, recounting their first jump, explained: 'I had an epiphany because I did not die but I really enjoyed it, a whole environment that I never imagined existed was opened to me. My life has been radically altered by that choice that day'.[47] The American mountaineer David Breashears described high-altitude climbing as providing 'a profound sense of self-knowledge'.[48]

One crucial benefit of extreme experiences is greater resilience.[49] As we saw in Chapter 11, coping successfully with stress can make us psychologically more resilient and better able to cope with future demands. Even relatively severe instances of stress can have good endings. Research has increasingly shown that traumatic experiences are not uniformly damaging and often have long-term positive effects—a phenomenon known as post-traumatic growth.[50]

In conclusion, we have seen that different psychological factors determine why an individual chooses to enter an extreme environment, why they stay there, and why they eventually leave. People stay for rewards such as mastery, commitment, and social support. Continued engagement may also be driven by a healthy or obsessive passion. People leave extreme activities for many other reasons, including disenchantment, declining capability, altered priorities, or changes in circumstances. The transition is likely to be harder for individuals with an obsessive passion, and those for whom the activity defines their social identity.

Lessons from hard places

Despite their dark sides, extreme environments have plenty of positive psychological effects. Most people emerge from them with greater resilience, a better understanding of their own strengths, and a sense of having lived life to the full. They derive satisfaction from having pushed themselves to the limits of their ability and mastering new skills. By facing the possibility of death, they become more appreciative of life and perhaps more humble. They learn to appreciate the value of things like good sleep and companionship.

Surviving and thriving in hard places requires focus—the ability to control our attention and attend wholly to a task or a feeling. The ability to focus underpins the development of expertise, helps in tolerating pain and hardship, and has long-term psychological and physical benefits.

For those of us who do not have the opportunity (or perhaps the inclination) to brave extreme environments, the lessons from hard places are just as relevant to softer places. Overcoming difficulties and achieving excellence requires careful planning and extensive preparation. This involves lots of work. Very few people get far on 'innate talent' alone. We could all benefit from cultivating focus, valuing sleep, being more tolerant and more tolerable, and accepting that some stress is good for us.

STUDYING THE PSYCHOLOGY OF EXTREME ENVIRONMENTS

Few people venture into extreme environments to work or play, let alone to carry out psychological research. Nevertheless, scientists have accumulated a substantial body of evidence about the psychology of people who choose extreme environments and how they cope once they get there.

The empirical evidence about the psychology of extreme environments falls into three broad categories: anecdotes, observational studies of people in actual and artificial extreme environments, and studies that attempt some measure of experimental control (field experiments). Together they provide a reasonably good picture of the most important issues. (Aspiring researchers should be reassured, however, that much remains to be discovered.) Each method has its own strengths and limitations, so scientists endeavour to develop an account that integrates evidence generated by a number of different methods. This multi-method approach, known as triangulation, reduces the bias that might be introduced by using only one method.

To give a flavour of how research is carried out in extreme environments, we will mention here some examples from the work of psychologists Peter Suedfeld and Jane Mocellin. Professor Suedfeld has made major contributions to the understanding of human behaviour in extreme environments and we have cited his research several times. Mocellin, now a distinguished expert in her own right, was Suedfeld's doctoral student in the 1980s. Both scientists have used a variety of research methods to study how people cope in isolated polar environments.

Anecdotes

If you have an interest in human endeavour then you may have read some of the substantial body of descriptive literature about extreme experiences. The public has an enduring fascination with the exploits of individuals who put themselves in harm's way and survive in dangerous environments that most of us could not tolerate. Bestselling books and popular films are testament to the appetite for real-life tales of heroism, endurance, and death-defying feats. As well as being entertaining, many of them convey a vivid sense of the problems people encounter and how they cope.

Beyond the adventure books, there exists an even richer body of data in the form of contemporaneous journals, letters, news reports, and weblogs. These provide detailed insights into people's experiences in extreme environments. Adventurers are often assiduous journal-keepers. Many expeditions involve long periods where little happens, and record keeping is a one way of filling the time. Solitude facilitates introspection, and recording their thoughts can help lone adventurers to deal with emotional challenges.[1] They also know that if for some reason they do not return to tell their story in person, their chronicle might be the only record of their exploits. Finally, sponsorship has always been an important part of expedition life, and sponsors, whether they are sixteenth-century monarchs or twenty-first-century corporations, demand records of the feats that they have made possible.

Such accounts are not in themselves scholarly research, but they can be a rich source of raw data. Archives have been created and preserved precisely because these raw data are intrinsically interesting, and the events they document are considered to be important.[2] These archives sometimes provide answers to questions that are too difficult to tackle in extreme environments using standard research techniques.

Anecdotal data have significant limitations, however. Accounts of expeditions inevitably leave out much more than they include, and, particularly when published, are likely to have been shaped to satisfy popular demands. Much of the mundane detail might be omitted, for instance, as might details that portray individuals (especially the writers) in an unattractive light. Hindsight can easily put a gloss on retrospective accounts, which also omit facts that the authors have genuinely forgotten or misremembered. The conclusions that can be drawn from research based solely on anecdotal accounts are therefore limited and tentative.

Despite these limitations, psychologists have derived new understanding from archival data. For example, Suedfeld and Mocellin examined historical accounts of polar expeditions to understand how early explorers coped with the hardships. Their analysis indicated that, far from emphasizing endless suffering, explorers tended to write more about positive experiences than negative ones. Indeed, the accounts were filled with 'appreciative comments concerning the grandeur and beauty of the natural surroundings; episodes of...serenity, and relaxation; and feelings of self-growth'.[3] Suedfeld later employed a similar method, this time drawing on astronauts' autobiographies, to explore the psychological impact of space travel.[4]

When analysed systematically and rigorously, archival records can provide rich descriptive accounts, sometimes leading to unexpected new insights. A thorough description, even one based on a single account, can be a starting point for further investigation and understanding.

Observational studies

Observational studies in real or artificial extreme environments are a more common source of empirical data. Researchers gather data from within the extreme environment itself or from the relative comfort of 'mission control', perhaps making brief visits to the extreme environment and interviewing and testing crewmembers in person. They deploy a range of research tools, from qualitative descriptions of behaviour to more systematic data gathering through the use of surveys, questionnaires, and interviews.

Studies of this type are logistically more complex and expensive to carry out than reading expedition logs, and they are therefore most likely to be government-sponsored. Observational studies often take place in facilities such as polar stations, submarine habitats, or space vessels.

In 'participant observation' studies, a scientist may join an expedition or other extreme mission specifically as a researcher, but they may be expected to carry out their research 'on the side'. For instance, the medical officer for the 1960 Australia National Antarctic Research Expedition was a working member of the fourteen-man expedition. He also kept a daily 'psychological log' recording the men's behaviour and emotions, which he later used as the basis for scholarly papers.[5]

In simulation studies, individuals volunteer to spend time in an environment that has been designed to replicate key features of the extreme environment. In the 1960s, NASA confined volunteers to Sealab II, an artificial underwater habitat designed to study the effects of prolonged confinement in dangerous conditions.[6] Groups of ten men at a time would spend two weeks 200 feet below the surface of the Pacific Ocean, carrying out oceanographic studies and going on 'sea walks'. They also completed numerous medical and psychological tests. Underwater habitats have been (and still are) run by agencies like NASA, who study how people cope in small, isolated teams, and test ways of carrying out fiddly mechanical repairs while in restrictive clothing and under dangerous conditions.

Another example is the Mars 500 project, in which six men spent 500 days confined in a capsule on a simulated mission to Mars. In reality, the 'space ship' was located in a car park in Moscow. Numerous studies were conducted throughout the simulation, to research the effects of prolonged confinement on everything from food palatability and sleep patterns to group cohesion and interpersonal conflict.[7]

People who join simulation studies know that the purpose is to research the effects on them of extreme environments. However, in other studies psychologists study individuals who are in extreme environments for quite different reasons—for instance, submarine crews, engineers in polar stations, or mountaineers on a summit attempt. This sort of research has yielded important findings but also raises practical and ethical concerns.

The first challenge is gaining permission even to attempt the research. Authorities or expedition leaders may deny a researcher access if they think the research will disrupt work or damage morale. If permission is granted, the researcher must then persuade people to participate in the research. The attitude of the potential subjects towards scientific research can range from enthusiastic interest to scepticism or even outright hostility. Few people enjoy being observed and measured while they are performing difficult tasks in demanding conditions. They may also worry that the results might affect their future employment prospects. This was a possible reason why, in one study of psychiatric issues in Antarctic winter-over crews, only a quarter of the civilians invited agreed to be debriefed by a mental health professional.[8]

Researchers then face three further hurdles: establishing credibility and trust with practitioners, who can be notoriously suspicious of outsiders; keeping the results confidential; and conducting meaningful research that causes minimum disruption to busy people's work.

Jane Mocellin and her colleagues faced these obstacles in their study of stress in polar environments.[9] The subjects were forty-seven people living and working in the Arctic (on two Canadian stations) and Antarctic (on two Argentine stations), plus another twenty-eight people working in the more comfortable Resolute Bay (in the Canadian Northwest Territories) and in Buenos Aires. The twenty-eight were a sort of control group, whom Mocellin and her colleagues compared to those working in the polar environments.

Mocellin spent two weeks as a participant observer in each place. Her first tasks were to establish her credibility, which she did by telling the crews about her previous polar experience, and reassuring them that the observations were not connected to any performance evaluations and would be confidential. As time went on, she joined in the regular parties that took place in the Arctic and Antarctic stations, drinking alongside crewmembers as a way of fitting in. She felt that she came to be accepted by these crews, who eventually spoke to her openly—and, she believed, genuinely—about their experiences and emotions. Mocellin had more difficulty with the crew at Resolute Bay, however, some of whom remained wary of her and refused to participate in the research.

In conventional research, psychologists ensure that individual participants in any study cannot be personally identified, and a breach of confidentiality would be considered an ethical problem. Maintaining participant confidentiality is easier when there are many subjects, as was the case in Mocellin's study. Confidentiality can be more problematic in cases where the identities of participants become public, either because the mission has been widely publicized or because the experiences described are extremely rare. For instance, the subject of a series of studies of a cosmonaut who spent 438 days in the Mir station can only be Valeri Polyakov (who, at the time of writing, still holds the record for the longest single space flight).[10] Scientists who wish to conduct research in extreme environments may need to consider from the outset whether their subjects are comfortable with the prospect of being identified.[11]

Relationships between teams and mission control sometimes become fraught, as we described in Chapter 8. One trigger for this can be the perceived insensitivity of researchers who are part of mission control. Whether observing first hand or from mission control, researchers must take care to ensure that their requests do not become overintrusive or harm participants' well-being. This may mean being alert to cues indicating that the participants resent being 'lab rats', but are not saying so for fear of being labelled as uncooperative.

In observational studies, psychologists gather data through behavioural observation, interviews, questionnaires, and surveys. Because people might deliberately or unconsciously exaggerate in self-report questionnaires to make themselves look better, researchers often use more than one data-gathering tool.[12] For a study that we described in Chapter 7, psychologist Gloria Leon investigated the experiences of members of an Arctic expedition. She did this by flying into their remote camp at intervals, collecting the questionnaires they had been filling in and carrying out lengthy one-to-one interviews.[13]

Hundreds of different questionnaires have been used to capture various measures of behaviour, experience, motivation, emotion, and cognition. Some questionnaires have been designed or adapted specifically for extreme environments. Most, however, are designed to measure universal psychological variables, such as personality traits or anxiety levels, which are equally relevant in extreme and everyday environments.

In some cases, such as when measuring stable personality traits, data can be gathered before or after the mission, when the participants are in normal environments. Sometimes researchers compare participants in different sorts of extreme or conventional activities— for example, skydivers versus mountaineers, or hang-gliders versus golfers—or compare experts with less experienced practitioners.[14] Such studies do not require psychologists to negotiate access to extreme environments—participants can, for instance, complete surveys online from the comfort of their own home. However, they rely on volunteers who are willing to complete questionnaires (and do so honestly), so researchers still face the challenge of establishing credibility with a potentially sceptical subject pool.

In other cases, participants are asked to complete questionnaires while they are in the extreme environment, perhaps because the researchers want to assess how it is affecting them at the time.[15] During Mocellin's polar study, she asked participants to complete several questionnaires. One of these was the well-established Spielberger State-Trait Anxiety Inventory (STAI).[16] State anxiety is a measure of an individual's (fluctuating) level of anxiety in response to specific situations, whereas trait anxiety is held to be a stable personality characteristic, related to how often and how intensely the individual feels anxious more generally.

Through these and other studies, Mocellin and Suedfeld found that people in isolated polar environments did not report high levels of trait anxiety. This is not very surprising, because natural worriers are unlikely to choose, or be chosen for, stressful missions. But neither were they particularly low in trait anxiety. In fact, the participants tended to score in the normal range. Another finding was that their state anxiety levels remained broadly stable through their entire stay, even though they were in situations that most of us would find highly stressful.[17] These results were consistent with findings from archival data. Together, they support the view that people who enter polar environments generally cope well, despite the hardships.

Field experiments

In a final category of data-gathering methods, which might be called field experiments, participants perform particular tasks in extreme environments so that scientists can measure the impact of specific factors on defined aspects of behaviour or cognition. Unlike conventional laboratory experiments, psychologists cannot study randomly assigned members of the public in extreme environments, and it is often impractical to have clearly defined 'treatment' and 'control' groups.[18] Instead, they study people who have already chosen (and who are prepared for) extreme activities. Most such studies compare each individual's performance at various points in the mission to see how it varies. (These are 'within-subjects' designs, in which each individual acts as their own control.)

In one study, for example, astronauts carried out tests of reasoning, memory, and fine motor control, before, during, and after spaceflight, in order to study the impact of microgravity on cognition.[19] Researchers working on Everest investigated the cognitive and linguistic deficits experienced by mountaineers using a similar design. The psychologists set up their lab at Everest Base Camp and administered tests of memory, speech, and fine motor control at various stages of the climb (including by radio link as the climbers ventured higher) to see what impact the lack of oxygen had on their cognition.[20]

Experiments are sometimes carried out in artificial settings that share crucial features with real extreme environments but which are cheaper and less dangerous. Examples include bed-rest studies to simulate aspects of microgravity, experiments in barometric chambers to explore the impact of changes in atmospheric pressure, and the sensory deprivation experiments we discussed in Chapter 5.

Psychologists have overcome major obstacles to investigate how people survive and thrive in extreme environments. They have designed imaginative and elegant studies, and conducted their investigations in environments that are not at all conducive to research. To achieve this, the psychologists themselves have required many of the qualities we have discussed in this book, including the ability to withstand hardship and work effectively with other people, know-how, resilience, and focus.

NOTES

Chapter 1

1. Cherry-Garrard (2010), p. xvii.
2. This was the eponymous journey in Cherry-Garrard's book *The Worst Journey in the World*, originally published in 1922.
3. Cherry-Garrard (2010), p. 265.
4. Cherry-Garrard (2010), pp. 242 and 304.
5. Macfarlane (2003).
6. Simpson (2003); Harrer (2005).
7. At the high pressure of depth, gases in the body dissolve into the blood and tissues; as the pressure reduces during the ascent, they become gases again. If the ascent is made too quickly bubbles form in the diver's tissues, causing nerve damage. Pausing on an ascent allows gases to be eliminated slowly and safely.
8. Finch (2008).
9. Finch (2008).
10. Obituary, *New York Times*, 10 October 2013.
11. Smith (2005).
12. Hamilton-Paterson (2010).
13. Hamilton-Paterson (2010).
14. Ryan (1995).
15. Ward and O'Brien (2007).
16. Roberts (2013), p. 72, quoting a member of Shackleton's 1908 British Antarctic Expedition.
17. Manzey and Lorenz (1998).
18. Lazarus (1966).
19. Lazarus and Folkman (1984); Hughes (2012).
20. Martin (1997).
21. Martin (1997).
22. Arnsten et al. (2012).
23. Cannon-Bowers and Salas (1998).
24. Martin (1997); Wetherell et al. (2006).
25. Dickerson and Kemeny (2004); Martin (1997); Wetherell et al. (2006).
26. Martin (1997).
27. Martin (1997); Wetherell et al. (2006); Dickerson and Kemeny (2004).
28. See for instance, Lazarus and Folkman (1984).
29. Lazarus and Folkman (1984).
30. Lazarus and Folkman (1984).
31. Martin (1997).
32. Lazarus and Folkman (1984); Cacioppo and Patrick (2008); Taylor (2011).

33. Palinkas et al. (2004a).
34. See for instance, Ritsher et al. (2007).
35. Martin (2005).

Chapter 2

1. Horace-Bénédict de Saussure (1740–1799), quoted in Macfarlane (2003), p. 71.
2. Richardson, in Brymer and Oades (2009), p. 122. BASE stands for Buildings, Antennas, Spans (bridges), and Earth (cliffs).
3. Cherry-Garrard (2010), p. 386.
4. Cherry-Garrard (2010), p. 396.
5. Everett (1891), pp. 41–2.
6. Rachman (1990).
7. Davis et al. (2010).
8. Rachman (1990); Wise (2009).
9. Wetherell et al. (2006); Colasanti et al. (2008); Rassovsky and Kushner (2003); Vickers et al. (2012).
10. Morgan (1995).
11. Simpson (1997), pp. 25–6.
12. Hall (2012).
13. McMurray (2001), p. 79.
14. Delle Fave et al. (2003).
15. Joseph (2011).
16. Gecas (1989).
17. Brymer and Oades (2009), p. 123.
18. Korchin and Ruff (1964), p. 205.
19. Ruff and Korchin (1964), p. 219.
20. Holahan et al. (2005); Lazarus and Folkman (1984).
21. Dias et al. (2012).
22. Castanier et al. (2011).
23. Van Schaik (2008), p. 18.
24. Coffey (2004), p. 12.
25. Barlow et al. (2013).
26. Rachman (1990), p. 12. See also Pury et al. (2010).
27. Rachman (1994).
28. Wise (2009).
29. Rachman (1990), drawing on the 1943 *Lancet* report of Sir Aubrey Lewis. See also Jones et al. (2004).
30. Askew and Field (2008).
31. Rachman (1990); McGurk and Castro (2010).
32. Rachman (1990).
33. Lester et al. (2010); Rachman (1990).
34. Knox-Johnston (2004).
35. Pury et al. (2010); Rachman (1990).
36. Franco et al. (2011).
37. Allison and Goethals (2010), p. 123.

38. Allison and Goethals (2010).
39. Fleming (2000).
40. Powter (2006), p. 51. See also Spufford (2003) and Davis (2012).
41. Huntford (1999). For a counter-view see, for instance, Fiennes (2013).
42. Arnett (1992); Duangpatra et al. (2009); Monasterio (2006).
43. Duangpatra et al. (2009); Teese and Bradley (2008).
44. Arnett (1995); Teese and Bradley (2008).
45. Duangpatra et al. (2009).
46. Gooderham (2009); Rapp (2008).

Chapter 3

1. Owen (2012), p. 69.
2. Thesiger (2007).
3. Fleming (2000).
4. Wilkinson (2012). This figure is based on an unreferenced journalistic report and should be treated with caution. Nevertheless, anyone familiar with the historical accounts will have little difficulty believing this figure, which is probably an underestimate.
5. Morris (1958), p. 89.
6. Krakauer (1997), p. 136.
7. Roach (2010), p. 233.
8. Fraser (1968).
9. Scott (2005), p. 26.
10. Roach (2010).
11. Mattoni and Sullivan (1962).
12. Suedfeld and Steel (2000).
13. Fiennes (2013), p. 346.
14. Nansen (2008), p. 528.
15. Fleming (2000).
16. Roberts (2013), p. 44.
17. Roberts (2013), p. 78.
18. Rose and Douglas (2000).
19. Curtis and Biran (2001); Oaten et al. (2009); Elwood and Olatunji (2009).
20. Curtis and Biran (2001).
21. Stuster (1996).
22. Greely (1886), p. 143.
23. Stuster (1996).
24. Fraser (1968), p. 42. See also Mattoni and Sullivan (1962); Roach (2010).
25. See for instance, Olatunji et al. (2007).
26. Stuster (1996).
27. Sontag and Drew (1999).
28. Ashcroft (2001). Fiennes (2013) states that at times during his two-man sledge-haul across the Antarctic he and his companion each expended over 10,000 calories per day.
29. Andrée et al. (1930); Broadbent (2007); Wilkinson (2012).
30. Roberts (2013).
31. Roberts (2013), p. 218.

32. Frost (2003).
33. Pemberton (2006).
34. Glouberman (2009).
35. Greely (1886), p. 135.
36. Perchonok and Douglas (2009).
37. Sandal et al. (2011).
38. Darlington (1957), p. 62.
39. Darlington (1957), p. 102.
40. Stein (2011), p. 402.
41. Greely, quoted in Stein (2011).
42. Guttridge (2000); Stein (2011).
43. Thornton (2010).
44. King (1878), in Nunn (1940).
45. Nunn (1940), pp. 357 and 363.
46. Nunn (1940), p. 358.
47. Rolls et al. (1980).
48. Thesiger (2007), p. 138.
49. These figures, which are the US official recommendations, are derived from median intakes rather than any universally agreed biological requirement.
50. Institute of Medicine (2004); Popkin et al. (2010); Thornton (2010).
51. Popkin et al. (2010).
52. Popkin et al. (2010); Neave et al. (2001); Rogers et al. (2001).
53. Cherry-Garrard (2010).
54. Fiennes (2013), pp. 308–15.
55. Jeal (2011).
56. Speke (1864).
57. Jeal (2011).
58. Cervero (2012), p. xii.
59. Wall (2000), p. 35. See also Bond and Simpson (2006); Bushnell et al. (2013); Cervero (2012); Martin (2008).
60. Cervero (2012); Bushnell et al. (2013).
61. Beecher (1946).
62. Cervero (2012).
63. Vachon-Pressau et al. (2013).
64. Cervero (2012).
65. Robinson et al. (2013).
66. Apkarian et al. (2005).
67. Yoshida et al. (2013).
68. Cervero (2012).
69. Cervero (2012); Bushnell et al. (2013).
70. Roberts (2013).
71. Kahneman et al. (1993).
72. Kahneman et al. (1993).
73. Wall (2000).
74. Karkshan et al. (2002); Minde (2006).
75. Cervero (2012), p. 8.

Chapter 4

1. Kanas and Manzey (2008); Barger et al. (2014).
2. Cherry-Garrard (2010).
3. Cherry-Garrard (2010), p. 291.
4. Thesiger (2007).
5. Buguet (2007); Johnson et al. (2010); Nussbaumer-Ochsner et al. (2012).
6. Roberts (2013).
7. Newby (1956), p. 74.
8. Newby (1956), p. 191.
9. Bennet (1973, 1983).
10. Bennet (1973).
11. Luyster et al. (2012); Martin (2002); Dement and Vaughan (1999); Pallesen et al. (2007).
12. Dijk et al. (2001); Santy et al. (1988).
13. Martin (2002).
14. Oswald (1974).
15. Lindbergh (1993), p. 232.
16. Lindbergh (1993).
17. Martin (2002).
18. Martin (2002).
19. Martin (2002); Harrison and Horne (2000); Durmer and Dinges (2005).
20. Lowden et al. (2010); Luyster et al. (2012); Cappuccio et al. (2011); Grandner et al. (2010).
21. Martin (2002); Dement and Vaughan (1999).
22. Tsai et al. (2005); Durmer and Dinges (2005); Martin (2002); Luyster et al. (2012).
23. Kanas and Manzey (2008).
24. Hurdiel et al. (in press).
25. Harrison and Horne (1999); Martin (2002).
26. Durmer and Dinges (2005); Connor et al. (2001); Williamson et al. (2011); Rajaratnam and Arendt (2001).
27. Ingre et al. (2006).
28. Blaivas et al. (2007); Martin (2002).
29. Dawson and Reid (1997); Williamson and Feyer (2000).
30. Martin (2002).
31. Martin (2002).
32. Hoeksema-van Orden et al. (1998); Shepperd (1993); Baranski et al. (2007).
33. Baranski et al. (2007).
34. Baranski et al. (2007).
35. Barnes and Hollenbeck (2009).
36. Barger et al. (2014).
37. Rajaratnam and Arendt (2001); Martin (2002); Åkerstedt (2003).
38. Imbernon et al. (1993).
39. Åkerstedt (2003).
40. Bjorvatn et al. (1998).
41. Rajaratnam and Arendt (2001); Martin (2002).
42. Martin (2002); Boggild and Knutsson (1999); Rajaratnam and Arendt (2001); Åkerstedt (2003).
43. Megdal et al. (2005); Schernhammer et al. (2001); Stevens (2009).
44. Rajaratnam and Arendt (2001); Caruso et al. (2004); Lowden et al. (2010).

45. Martin (2002).
46. Buguet (2007).
47. Martin (2002).
48. Martin (2002).
49. Martin (2002); Milner and Cote (2009); Hayashi et al. (1999); Mednick et al. (2002); Takahashi and Arito (2000).
50. Martin (2002); Rajaratnam and Arendt (2001); Åkerstedt (2003).
51. Asaoka et al. (2012).
52. Hofer-Tinguely et al. (2005); Signal et al. (2012).
53. Antonenko et al. (2013); Lau et al. (2010).
54. Faraut et al. (2011); Vgontzas et al. (2007).
55. Martin (2002); Rihel and Schier (2013).
56. Lorist and Tops (2003).
57. Martin (2002).
58. Mednick et al. (2008).
59. Kanas and Manzey (2008).
60. Fiennes (2013), pp. 245–6.
61. Mignot (2013).
62. Knox-Johnston (2004), p. 97.
63. Martin (2002).

Chapter 5

1. Nansen (2008).
2. Siffre (1964, 1975).
3. Siffre (1975), p. 428.
4. Siffre (1975), p. 431.
5. Siffre (1975), p. 435.
6. *Los Angeles Times*, 9 December 1988.
7. Bexton et al. (1954).
8. Bexton et al. (1954), p. 74.
9. Zuckerman (1969).
10. Suedfeld and Coren (1989).
11. James (1890), p. 759.
12. Kumar et al. (2009).
13. Sacks (2012); Blanke et al. (2008).
14. Hill and Linden (2013).
15. Ohayon (2000).
16. Hill and Linden (2013).
17. Teunisse et al. (1995, 1996); Sacks (2012).
18. McCorristine (2010).
19. Brugger et al. (1999).
20. Siegel (1984).
21. Suedfeld and Mocellin (1987).
22. Blanke et al. (2008).
23. Geiger (2009).
24. Geiger (2009).

25. Brugger (2006).
26. Booth et al. (2005).
27. Cheyne (2009) argues that sensed presence experiences are delusions rather than hallucinations. For the sake of simplicity we have conflated the two.
28. Bennet (1983), p. 143.
29. Bentall (1990); Cheyne (2012); Chan and Rossor (2002).
30. Suedfeld and Bow (1999); Van Dierendonck and Te Nijenhuis (2005); Bood et al. (2006).
31. Van Schaik (2008), p. 124.
32. Finch (2008).
33. Thesiger (2007), p. 177.
34. Owen (2012).
35. Delle Fave et al. (2003).
36. Cernan and Davis (1999), p. 310.
37. Kanas and Manzey (2008); Roach (2010).
38. Martin (2008).
39. Suedfeld and Steel (2000).
40. Suedfeld and Steel (2000).
41. Dumas (2011).
42. Stuster (1996); Suedfeld and Steel (2000).
43. Greely (1886), p. 142.
44. Roberts (2013).
45. Stuster (1996).
46. Hadfield (2013).
47. Corbett (2004).
48. Griffiths (2007), p. 176.
49. Stuster (1996).
50. Guly (2013).
51. Suedfeld and Steel (2000); Bhargava et al. (2000); Cravalho (1996).
52. Martin (2008).
53. Wheeler (1997).
54. Stuster (1996).
55. Cook (1900), p. 296.
56. Stuster (1996).
57. Darlington (1957).
58. Cook (1900), p. 303.
59. Thesiger (2007), p. 158.
60. Yeomans (1998).
61. Ferguson (1970).
62. Remick et al. (2009).
63. Stuster (1996).
64. Hadfield (2013).
65. Zorpette (1997), p. 36.
66. Karafantis (2013).
67. Yan and England (2001); Kanas and Manzey (2008); Peldszus et al. (2014).
68. Slack et al. (2009).
69. Kanas and Manzey (2008).
70. Kanas and Manzey (2008).

71. Weybrew (1991).
72. Stuster (1996); Clearwater and Coss (1991).

Chapter 6

1. Byrd (1938), p. 4.
2. Roberts (2013), pp. 198–200.
3. Knox-Johnston (2004).
4. Slocum (2006), p. 31.
5. Hames et al. (2013).
6. Silvia and Kwapil (2011).
7. Kwapil (1998); Silvia and Kwapil (2011).
8. Hills and Argyle (2001); Coplan and Weeks (2010a).
9. Leary et al. (2003a).
10. Burger (1995).
11. Cramer and Lake (1998).
12. Long et al. (2003).
13. Storr (1988). See also Rufus (2003).
14. Coplan et al. (2004); Coplan and Weeks (2010a, 2010b).
15. Coplan et al. (2004); Rubin et al. (2009).
16. Thesiger (2007), p. 120.
17. Peplau and Perlman (1982); Russell et al. (2012).
18. Hawkley et al. (2003).
19. Perlman and Peplau (1984). You can measure your current state of loneliness using the UCLA Loneliness Scale: Russell (1996).
20. Cacioppo and Patrick (2008); Martin (1997); Jaremka et al. (2013).
21. Downey et al. (1998); Heinrich and Gullone (2006); Segrin and Kinney (1995).
22. Byrd (1938), p. 197.
23. Kwapil (1998); Silvia and Kwapil (2011).
24. Buchanan (2000); Powter (2006); Krakauer (1997).
25. Shalev (2008); Grassian (1983).
26. Arrigo and Bullock (2008); Scott and Gendreau (1969).
27. 'A voyage for madmen' is the title of the account by Nichols (1997).
28. Nichols (1997); Tomalin and Hall (2003).
29. Nichols (1997); Tomalin and Hall (2003); Bennet (1974); Moitessier (1995).
30. Bennet (1974, 1983); Powter (2006).
31. Kull (2008), p. 62.
32. Storr (1988).
33. Maitland (2008), p. 48.
34. Maitland (2008), pp .66–74.
35. Lester (1983), p. 38. See also, Brymer and Gray (2010); Maitland (2008); Lester (2004).
36. Simons and Schanche (1960), quoted in Ryan (1995).
37. Clark and Graybiel (1957), pp. 122–3.
38. Courtauld (1932), p. 73.
39. Zubek (1973).
40. Milnes Walker (1972), pp. 51–2.
41. Zubek (1973).

42. Haggard (1973).
43. Knox-Johnston (2004).
44. Baumeister and Leary (1995).
45. James (1978); Knox-Johnston (2004); Milnes Walker (1972); Slocum (2006).
46. Siffre (1975).
47. Mills (1964).

· Chapter 7

1. Stuster (1996), p. 165.
2. Thesiger (2007), p. 52.
3. Kanas (1998).
4. Byrd (1938), p. 16.
5. Wheeler (1997).
6. McCormick et al. (1985); Rivolier et al. (1988, 1991); Taylor (1991).
7. Rivolier et al. (1991). The authors tactfully omit to say which nationality. Your choices are: Argentina, France, New Zealand, Australia, or Britain.
8. Bates and Plog (1976).
9. Suedfeld et al. (2011). We are using 'astronaut' as a catchall term that includes Russian cosmonauts.
10. Suedfeld et al. (2011).
11. Suedfeld et al. (2011).
12. Inoue et al. (2004); Sandal (2004); Ritsher (2005).
13. Inoue et al. (2004), p. C33.
14. Suedfeld et al. (2011), p. 151.
15. Frost, (1995), p. 133.
16. Finney (1991).
17. Griffiths (2007), p. 178.
18. Franklin (2011).
19. Palmai (1963).
20. Gruter and Masters (1986).
21. Palmai (1963).
22. Griffiths (2007).
23. Williams (2007).
24. Williams (2007); Nezlek et al. (2012); Williams and Nida (2011).
25. Williams and Nida (2011).
26. Williams et al. (2000); Smith and Williams (2004).
27. Gonsalkorale and Williams (2007).
28. Eisenberger et al. (2003). Ostracism and physical pain are both associated with activation in the dorsal anterior cingulate cortex.
29. Baumeister et al. (2002, 2006).
30. Leary et al. (2003b).
31. Williams and Nida (2011); Williams (2007).
32. Williams and Sommer (1997).
33. Oaten et al. (2008); Nezlek et al. (2012); Williams (2007); Twenge et al. (2001).
34. Downey et al. (1998).
35. Palinkas (1991).

36. Altman (1975); Leino-Kilpi et al. (2001).
37. Stuster (1996).
38. Altman (1975, 1976).
39. Altman (1976); Evans et al. (2000).
40. Thesiger (2007).
41. Letter to J.S. Henslow, 15 November 1831, <http://www.darwinproject.ac.uk/letter/entry-147> accessed February 2014.
42. Stuster (1996).
43. Kanas and Manzey (2008); Yan and England (2001).
44. Suedfeld and Steel (2000).
45. See for instance, Oberg and Oberg's (1986) account of the practical difficulties of sex in zero gravity.
46. Wetzler (2001).
47. Tabor (2011), p. 151.
48. Stuster (1996); Fleming (2000).
49. Burns (2001); Conefrey (2011).
50. Wetzler (2001).
51. Tabor (2011), p. 151.
52. Burns (2001).
53. Coffey (2004), p. 74.
54. Cherry-Garrard (2010), p. 596.
55. Palmai (1963), pp. 364–7.
56. Cravalho (1996), p. 152.
57. Suedfeld and Steel (2000).
58. Fiennes (2013), p. 135.
59. Roach (2010), p. 36.
60. Mishra (2006).
61. Bennet (1973).
62. Knox-Johnston (2004).
63. Stuster (1996).
64. Thesiger (2007), p. xii.
65. Burgess et al. (2001); Donnelly et al. (2001).
66. Dutton and Aron (1974) is still widely cited for the 'Hollywood effect'.
67. Fiennes (2007).
68. Darlington (1957).
69. O'Brien (1999).
70. Leon et al. (2002).
71. Leon et al. (2002); Leon and Sandal (2003). The Sverdrup 2000 Expedition website can be accessed via <http://web.archive.org/web/20011202030032/http://www.sverdrup2000.org/index.htm> accessed February 2014.
72. Landreth (2003).
73. Leon et al. (2002).
74. Wu (2013).
75. Coffey (2004).
76. Macfarlane (2003), p. 98.
77. Coffey (2004), p. 35.
78. Coffey (2004).

79. N. Lawson, *The Times*, 16 May 1995.
80. Rose and Douglas (2000).
81. Davis (2012).
82. Wilkinson (2012), p. 57.
83. 2008 documentary *Solo (Solitary Endeavour on the Southern Sea)*.

Chapter 8

1. Bishop et al. (2000).
2. Kozlowski and Ilgen (2006).
3. Kozlowski and Bell (2003).
4. Markman and Gentner (2001).
5. DeChurch and Mesmer-Magnus (2010); Kozlowski and Ilgen (2006); Langan-Fox et al. (2001).
6. Kozlowski and Ilgen (2006); Collyer and Malecki (1998). Langan-Fox et al. (2001) argue that a team mental model is not the same as a shared mental model. Strictly speaking they are right, but for the sake of simplicity we are using the terms interchangeably.
7. Kozlowski and Ilgen (2006); Cannon-Bowers and Salas (1998).
8. Kozlowski and Ilgen (2006).
9. Bishop et al. (2000).
10. Kozlowski and Ilgen (2006).
11. Tumbat and Belk (2011).
12. Dion (2004).
13. Beal et al. (2003).
14. Kozlowski and Ilgen (2006).
15. Stajkovic and Luthans (1998); Gully et al. (2002).
16. Hogg (1993).
17. Kelly and Kanas (1992).
18. Schneider and Reichers (1983).
19. Hogg (1993).
20. The classic account of groupthink is Janis and Mann (1977).
21. Orasanu (2005).
22. Orasanu (2005).
23. Radloff and Helmreich (1968).
24. Sexton and Helmreich (2000); Orasanu (2010).
25. Lebedev (1988), quoted in Kring and Kaminski (2011), p. 131.
26. Rosnet et al. (2004).
27. Kring and Kaminski (2011).
28. Leon and Sandal (2003).
29. Blum (2005); Conefrey (2011).
30. Atlis et al. (2004); Kahn and Leon (1994); Leon and Sandal (2003).
31. Arnesen et al. (2004); Atlis et al. (2004).
32. Leon and Sandal (2003).
33. Leon (2005).
34. Leon (2005).
35. Leon and Sandal (2003).
36. Hersch (2012), p. 32.

37. Ruff (2010), p. 157.
38. Korchin and Ruff (1964); Ruff and Korchin (1964).
39. Ruff and Korchin (1964).
40. Korchin and Ruff (1964).
41. Santy (1994).
42. Hylton (2007).
43. Wolfe (2005), p. 378.
44. See for instance, Dion (2004).
45. Hadfield (2013), p. 102.
46. Kanas and Manzey (2008).
47. Morrell and Capparell (2003).
48. Shackleton (1920).
49. Driskell et al. (2006).
50. Totterdell et al. (1998).
51. Driskell et al. (2006); Dion (2004).
52. Burke et al., (2006); Kozlowski, et al. (1999); Kozlowski et al. (2009); Serfaty et al. (1998).
53. Yamagishi (2001); McKnight et al. (1998).
54. Dirks (2000).
55. De Jong and Dirks (2012).
56. Sarris (2007).
57. Shuffler et al. (2011).
58. Shuffler et al. (2011).
59. Suedfeld et al. (2011).
60. Salas et al. (2012).
61. See for instance, the first-hand account of the Mars 500 simulation study, in which six men were isolated for 500 days on a mock mission to Mars, in Urbina and Charles (2014).
62. See <http://www.esa.int> accessed February 2014.
63. Stogdill (1974), p. 259.
64. Cherry-Garrard (2010), pp. 205–7.
65. Stuster (1996).
66. Byrd and Gould (1930).
67. Hannah et al. (2010).
68. Kolditz (2007), p. 70.
69. Hovland et al. (1953); Posner and Kouzes (1988); Kouzes and Posner (1992).
70. Kolditz (2007).
71. Dirks and Ferrin (2002).
72. Kolditz (2007).
73. Einarsen et al. (2007); Padilla et al. (2007); Shaw et al. (2011).
74. Reed (2004); Steele (2011).
75. Hogan (2006).
76. Padilla et al. (2007).
77. Lipman-Blumen (2005).
78. Gilbert et al. (2012); Bennis (2007).
79. For instance, Johnson (2010).
80. Kelly and Kanas (1992); Palinkas et al. (2000).
81. Slack et al. (2009).
82. Moskovitz (2011).

83. Kanas and Manzey (2008).
84. Vaernes (1993); Kanas et al. (1996).
85. Kanas et al. (2007).
86. Kanas et al. (2001).
87. Franklin (2011), p. 206.
88. Chaikin (1994); Reason (2008).
89. Kanas and Manzey (2008).

Chapter 9

1. Finch (2008); Mitchell et al. (2007).
2. Hadfield (2013), p. 65.
3. Simpson (1997).
4. Orasanu (1997).
5. Orasanu (1997, 2005).
6. Colvard and Colvard (2003); Morgan (1995).
7. Casteret (1938), p. 119.
8. Hoffman (2001).
9. Nelson et al. (1990); Kamler (2004); Ashcroft (2001).
10. Nelson et al. (1990).
11. Ryan (1995), p. 92.
12. Petri (2003); Piantadosi (2003).
13. Kanas and Manzey (2008).
14. Kanas and Manzey (2008).
15. Sicard et al. (2001).
16. Acheson et al. (2007); Drummond et al. (2006); Durmer and Dinges (2005); Killgore (2007); McKenna et al. (2007); Martin (2002).
17. Drummond et al. (2006).
18. Acheson et al. (2007).
19. Kuhnen and Knutson (2005). The brain regions involved were the nucleus accumbens (risk seeking) and anterior insula (risk aversion).
20. Burke et al. (2008).
21. Helweg-Larsen and Shepperd (2001); Shepperd et al. (2013). Optimism bias for a specific context should not be confused with optimism as a personality trait, see Radcliffe and Klein (2002).
22. Arnett (1995).
23. Martha and Laurendeau (2010)
24. Ben-Zur and Zeidner (2009).
25. See for instance, Ewert (1994).
26. Plous (1993).
27. Hoorens (1993); Dunning et al. (2003). For over-optimism in drivers, see Svenson (1981); in doctors, see Hodges et al. (2001); and in professors, see Cross (1977).
28. Simonet and Wilde (1997); Adams (2013).
29. Skydiver Bill Booth claimed this, tongue-in-cheek, as 'Booth's Law #2' in 2003: <http://www.dropzone.com/cgi-bin/forum/gforum.cgi?post=538444> accessed February 2014.
30. Stanton and Pinto (2000); Jackson and Blackman (1994); Viscusi (1984).
31. Lasenby-Lessard and Morrongiello (2011).

32. Wilde et al. (2002).
33. Kayes (2006).
34. Baron (2000).
35. Cialdini and Goldstein (2004).
36. Bowley (2010); Viesturs and Roberts (2009).
37. Bowley (2010); Viesturs and Roberts (2009).
38. Chi et al. (1988); Klein (1997).
39. Klein (1997, 1998).
40. Lipshitz et al. (2001).
41. Endsley (1995).
42. Dreyfus (1997); Endsley (1995).
43. Flanagan (1954); Klein (1998).
44. Klein (1998).
45. Schoenfeld (1992), p. 51.
46. Endsley (2000); Lipshitz and Shaul (1997); Markman and Gentner (2001).
47. Klein and Crandall (1995).
48. Viesturs and Roberts (2006), p. 311.
49. Westhoff et al. (2012).
50. Wheeler (1997), p. 103.
51. Fiennes (2013).
52. Tabor (2011).
53. Stone (2004), p. 72.
54. Stone (2004), p. 73.
55. Lindbergh (1993).
56. Kahneman and Tversky (1982).
57. Buehler et al. (1994); Sanna et al. (2005); Flyvbjerg et al. (2009).
58. Kahneman and Tversky (1982); Newby-Clark et al. (2000).
59. Cohen et al. (1997).
60. Hadfield (2013), p. 61.
61. Bown (2012), p. 8.
62. Burtscher (2012).
63. Ericsson (2006); Ericsson and Charness (1994).
64. Yates (2001).
65. Duckworth et al. (2007), p. 1088.
66. Duckworth et al. (2007).
67. Johnson et al. (2006).
68. Ericsson and Lehman (1996).
69. Johnson et al. (2006), p. 132.
70. Meichenbaum and Deffenbacher (1988).
71. Mocellin (1995). Mocellin did note that there may have been some 'social desirability' effect in her findings: any admission of weakness might have been damaging for these elite officers' careers.
72. See for instance, Hadfield (2013).
73. Whiteclay Chambers III (2010), pp. 6–7.
74. Simpson (1997).
75. Baltes and Staudinger (2000); Sternberg (2001); Brugman (2006); Meeks and Jeste (2009).
76. Ardelt (2003); Jeste et al. (2010).

77. Jeste et al. (2010).
78. Linley (2003).

Chapter 10

1. CBS 60 Minutes documentary *The Ascent of Alex Honnold*, broadcast 2 October 2011, available at <http://www.cbsnews.com/video/watch/?id=7383158n> accessed February 2014, and *Alone on the Wall* (Sender Films, 2010).
2. Loewenstein (1999), p. 333.
3. Lester (1983), pp. 40–1.
4. Cherry-Garrard (2010), p. 247.
5. Simpson (1997).
6. Kjærgaard et al. (in press).
7. Smallwood et al. (2008); Mason et al. (2007).
8. Smallwood and Schooler (2006); Baird et al. (2011).
9. Carriere et al. (2008).
10. Smallwood et al. (2008).
11. Baird et al. (2011).
12. Baird et al. (2011); Bateson and Martin (2013).
13. Morton (2003), p. 45.
14. Finch (2008); McMurray (2001); Van Schaik (2008).
15. Burroughs (1999).
16. Suedfeld and Steel (2000).
17. Kjellberg et al. (1998); Abbate et al. (2004); Ljungberg and Parmentier (2010); Szalma and Hancock (2011).
18. Weinstein (1978).
19. Smith (2012).
20. Martin (1997); Smith, A.P. (2012); Bucks et al. (2008).
21. Smallwood et al. (2009).
22. Cherry-Garrard (2010), p. 212.
23. Fitch and Ravlin (2005); Bénabou and Tirole (2004).
24. Mischel et al. (2011).
25. Gailliot et al. (2007).
26. Baumeister et al. (1998).
27. Gailliot et al. (2007).
28. Gailliot et al. (2007); Masicampo and Baumeister (2008).
29. Mischel et al. (2011); Bénabou and Tirole (2004).
30. Csikszentmihalyi (1990), p. 4.
31. Csikszentmihalyi (1990); Jackson and Eklund (2004).
32. Jackson and Csikszentmihalyi (1999); Jackson et al. (2001).
33. Logan (1985); Zorpette (1999).
34. Hunter (2011), p. 182.
35. Finch (2008). See also Delle Fave et al. (2003); Lester (1983).
36. Chen and Chen (2011).
37. Mathwick and Rigdon (2004); Mauri et al. (2011).
38. Martin (2005); Ceja and Navarro (2011).
39. Coffey (2004); Maass (2002).

40. Kabat-Zinn (1982); Cahn and Polich (2006); Miller et al. (1995); Grossman et al. (2004).
41. Kabat-Zinn (1994), p. 4.
42. Lutz et al. (2009).
43. Jha et al. (2007); Lutz et al. (2009); Travis et al. (2002).
44. Zeidan et al. (2010a).
45. Cahn and Polich (2006); Lazar et al. (2005).
46. Brefczynski-Lewis et al. (2007).
47. Lazar et al. (2005).
48. See also Kaplan (2001).
49. Berto (2005). See also Kaplan (1995, 2001); Ulrich et al. (1991).
50. Fisher (1993), p. 396. See also Pekrun et al. (2010); Tilburg and Igou (2012).
51. Martin (2008); Spacks (1995).
52. Eastwood et al. (2012); Svendsen (2005); Tilburg and Igou (2012).
53. Carriere et al. (2008); Eastwood et al. (2012); Fisher (1993); Pekrun et al. (2010); Seib and Vodanovich (1998).
54. Critcher and Gilovich (2010).
55. Fisher (1998).
56. Damrad-Frye and Laird (1989).
57. Vodanovich (2003).
58. Blaszczynski et al. (1990); Carlson et al. (2010); Lee et al. (2007); Mercer and Eastwood (2010); Vodanovich (2003).
59. Watt and Vodanovich (1999).
60. Kass et al. (2001).
61. Kass et al. (2001); Harris (2000).
62. Culp (2006); Zuckerman (2006).
63. Bartone and Adler (1999).
64. Sommers and Vodanovich (2000); Vodanovich (2003); Goldberg et al. (2011).
65. Britton and Shipley (2010).
66. Seib and Vodanovich (1998); Trunnell et al. (1996).

Chapter 11

1. Ritsher et al. (2007).
2. See <http://www.apa.org/helpcenter/road-resilience.aspx#> accessed February 2014.
3. Bonanno (2004); Connor and Davidson (2003); Feder et al. (2009); Martin (2005); Southwick and Charney (2012).
4. Southwick and Charney (2012); Joseph (2011).
5. Connor and Davidson (2003).
6. Bonanno (2004); Bartone (1999).
7. Dion (2004); Maddi and Khoshaba (1994).
8. Bartone (1999).
9. Crust (2008); Jones et al. (2002); Sheard (2010).
10. Crust and Swann (2013).
11. Crust and Clough (2005).
12. Kanas and Manzey (2008); Sandoval et al. (2012); Slack et al. (2009).
13. Palinkas et al. (2004b).
14. Décamps and Rosnet (2005).

15. Cravalho (1996).
16. Maslach and Leiter (1997); Leiter et al. (2010).
17. Hughes (2012); Cohen et al. (2013).
18. Fear et al. (2010).
19. Galea et al. (2003); Bonanno et al. (2006).
20. Bonanno (2005); Hughes (2012).
21. Horsburgh et al. (2009).
22. Binder et al. (2008); Hughes (2012).
23. Hughes (2012).
24. Morgan et al. (2000); Feder et al. (2009).
25. Hughes (2012).
26. King et al. (1998).
27. Southwick and Charney (2012).
28. Frankl (1984).
29. Bonanno (2004); Martin (2005).
30. Martin (2001).
31. Bateson and Martin (2013).
32. Peldszus et al. (2014).
33. Sontag and Drew (1999).
34. Linenger (2000), quoted in Slack et al. (2009), p. 13.
35. Wheeler (1997).
36. Keenan (1993), p. 295.
37. Keenan (1993), pp. 246 and xvi.
38. Martin (2001); Martin (2005).
39. Fredrickson et al. (2003).
40. Bonanno (2005).
41. Cahn and Polich (2006); Davidson et al. (2003); Grossman et al. (2004); Carlson et al. (2004); Kabat-Zinn (2003); Tang et al. (2007); Jha et al. (2007).
42. Miller et al. (1995).
43. Grossman et al. (2004).
44. Ramel et al. (2004).
45. Jung et al. (2010).
46. Tang et al. (2007).
47. Jha et al. (2007); Grossman et al. (2004); Zeidan et al. (2010b, 2011).
48. Kabat-Zinn (1982).
49. Kabat-Zinn et al. (1985).
50. Zeidan et al. (2010b, 2011).
51. Gard et al. (2012).
52. Davidson et al. (2003).
53. Dusek et al. (2008); Bhasin et al. (2013).
54. Speca et al. (2000).
55. Cervero (2012).
56. Stuster (1996).
57. Amundsen (1913), quoted in Guly (2013), p. 95.
58. Mountfield (1974), cited in Stuster (1996).
59. Darlington (1957), p. 157.

60. A BBC *Horizon* documentary about the scientists, entitled 'Inside Chernobyl's Sarcopha-gus', can be found on You Tube: <http://www.youtube.com/watch?v=h437B0YHKPs> accessed February 2014.
61. Kanas and Manzey (2008).
62. Yerkes and Dodson (1908).
63. Levine (1957, 2005).
64. Seery (2011); Seery et al. (2013).
65. Lyons and Parker (2007); Hughes (2012); Lewitus and Schwartz (2009).
66. Gouin et al. (2008); Aschbacher et al. (2013).
67. Elder and Clipp (1989).
68. Taleb (2012), p. 3.
69. McCay et al. (1935).
70. Lee and Longo (2011); Colman et al. (2009).
71. Poynter (2006); Alling et al. (1993); Reider (2010).
72. Walford et al. (2002).
73. Wan et al. (2010); Mattson and Wan (2005); Stote et al. (2007); Green et al. (2011); Lee and Longo (2011).
74. Green et al. (2011); Halagappa et al. (2007); Mattson (2008).
75. Mattson and Wan (2005); Mattson (2008).
76. Lee and Longo (2011); Green et al. (2011).

Chapter 12

1. Kittinger and Ryan (2011); Ryan (1995).
2. For more on pleasure versus satisfaction, see Martin (2005).
3. Pervin and John (1997).
4. Nettle (2007); John et al. (2008).
5. Nettle (2007); Cain (2012).
6. Zuckerman (1994), p. 27.
7. Dahlen and White (2006); Wagner (2001).
8. Cazenave et al. (2007).
9. Wheeler, in the introduction to Cherry-Garrard (2010), p. xiii.
10. Hale (2009).
11. Hale (2009), p. 148.
12. Wagner and Houlihan (1994); Diehm and Armatas (2004).
13. Leon et al. (1989).
14. Cronin (1991).
15. Barlow et al. (2013).
16. Rowland et al. (1986).
17. Bateson and Martin (2013).
18. Thesiger (2007).
19. Wilkinson (2012).
20. From the 1965 Foreword by George Seaver to Cherry-Garrard (2010).
21. MacKersey (1990).
22. MacKersey (1990); the psychologist is Powter (2006).
23. Schoonover (2007).
24. Martin (2008).

25. Churton (2011); Powter (2006).
26. Kittinger and Ryan (2011).
27. Smith (2002).
28. Bateson and Martin (1999); Bright et al. (2009); Krantz (1998).
29. Darwin (1958), p. 28.
30. Noyce (1958).
31. *Independent*, 25 August 2013.
32. Lopes Cardozo et al. (2012).
33. Darwin (1958).
34. Boston (2004). See also <http://www.ted.com/talks/penelope_boston.html> accessed February 2014.
35. Cabrol (2004), p. 62.
36. Norton (2000), p. 126.
37. Svendsen (2005), p. 141.
38. Svendsen (2005); Tilburg and Igou (2012).
39. Sarramon et al. (1999); Watt and Vodanovich (1992); Todman (2003).
40. Todman (2003); Dahlen et al. (2005); Ferrell (2004).
41. Russell (1950), p. 4.
42. Martin (2008).
43. Macintyre (2007).
44. Macintyre (2007), p. 134.
45. Ryan (2003).
46. Ryan (2003), p. 41.
47. Piantanida's gondola is exhibited in the Smithsonian National Air and Space Museum. It still carries the handmade sign 'JADODIDE', representing the names of his wife and daughters: Janice, Donna, Diane, and Debbie.
48. Ryan (2003), p. 258.
49. Ryan (2003), p. 23.

Chapter 13

1. Tajfel (1982); Ellemers et al. (2002).
2. Macfarlane (2003), p. 225.
3. Macfarlane (2003), p. 272.
4. Vallerand (2008), pp. 1–2.
5. Vallerand (2012).
6. Vallerand et al. (2003).
7. Vallerand (2008, 2012); Vallerand et al. (2003); Bélanger et al. (2013).
8. Bélanger et al. (2013).
9. Bélanger et al. (2013).
10. Vallerand (2012), p. 9.
11. Amiot et al. (2006).
12. Vallerand (2012).
13. Vallerand (2012).
14. Vallerand (2012).
15. Rip et al. (2006).
16. Vallerand et al. (2003).

17. Amiot et al. (2006).
18. Cialdini and Goldstein (2004).
19. Ryan (2003).
20. Powter (2006).
21. Meyer and Allen (1991); Meyer et al. (2002).
22. Ebaugh (1988).
23. Aldrin (2009); Smith (2005).
24. Aldrin (2009), p. 59.
25. Smith (2005).
26. Garrett E. Reisman, in the *New York Times*, 23 April 2011.
27. Martin (1997); Wanberg (2012).
28. Wanberg (2012).
29. Pritchard (1999, 2005). See also <http://ppritchard.blogspot.co.uk> accessed February 2014.
30. Kerr (2007), p. 346.
31. Coffey (2004); Wickwire and Bullitt (1998).
32. Macfarlane (2003), p. 98.
33. See <http://www.archives.gov/press/press-kits/american-originals-photos/moon-disaster-1.jpg> and <http://www.archives.gov/press/press-kits/american-originals-photos/moon-disaster-2.jpg> accessed February 2014.
34. Abramovitvch (2005).
35. Andrée et al. (1930), p. 388.
36. Wilkinson (2012).
37. Boss (1999, 2002); Frankl (1984).
38. Coffey (2004), p. 129.
39. Coffey (2004).
40. Palinkas and Suedfeld (2008); Suedfeld et al. (2011).
41. Mocellin and Suedfeld (1991).
42. Kjærgaard et al. (2013); Kjærgaard et al. (in press).
43. Palinkas (1991).
44. Suedfeld and Steel (2000); Suedfeld and Weiszbeck (2004); Suedfeld (2010).
45. Mitchell (2008), pp. 4–5. See also Hadfield (2013) and Harrison (2001).
46. Brymer and Schweitzer (2013).
47. Brymer and Oades (2009), p. 118.
48. Breashears (2000), p. 305.
49. Suedfeld et al. (2011).
50. Joseph (2011); Tedeschi and Calhoun (2004).

Appendix

1. Pennebaker (1997).
2. Simonton (2003).
3. Mocellin and Suedfeld (1991), p. 716.
4. Suedfeld and Weiszbeck (2004). For other examples, see Guly (2013); Peldszus et al. (2014); Stuster (1996, 2010); Stuster et al. (1999).
5. Palmai (1963). Other examples include Cravalho (1996); Brymer and Oades (2009).
6. Karafantis (2013); Hellwarth (2012).

7. Urbina and Charles (2014).
8. Palinkas et al. (2004b).
9. Mocellin et al. (1991).
10. Manzey et al. (1998).
11. Leon (2005); Ritsher et al. (2005).
12. Décamps and Rosnet (2005).
13. Leon et al. (2002).
14. For more on skydivers versus mountaineers, see Barlow et al. (2013); on hang-gliders versus golfers, see Wagner and Houlihan (1994); and on experts versus less experienced practitioners, see Martha and Laurendeau (2010).
15. Another example is Egan and Stelmack (2003).
16. Spielberger (1972); Barnes et al. (2002).
17. Mocellin et al. (1991).
18. For a general account of experimental design and observational methodology, see Martin and Bateson (2007) and Robson (2011).
19. Kanas and Manzey (2008).
20. Lieberman et al. (2005).

REFERENCES

Abbate, C. et al. (2004). Affective correlates of occupational exposure to whole-body vibration: A case-control study. *Psychotherapy Psychosomatics, 73,* 375–9.

Abramovitvch, H. (2005). Where are the dead? Bad death, the missing, and the inability to mourn. In S.C. Heilman (Ed.) *Death, bereavement, and mourning* (pp. 53–67). New Jersey: Transaction Books.

Acheson, A., Richards, J.B., and de Wit, H. (2007). Effects of sleep deprivation on impulsive behaviors in men and women. *Physiology and Behavior, 91,* 579–87.

Adams, J. (2013). Risk compensation in cities at risk. In T. Rossetto et al. (Eds), *Cities at risk: Living with perils in the 21st century, advances in natural and technological hazards research 33* (pp. 25–44). Netherlands: Springer.

Åkerstedt, T. (2003). Shift work and disturbed sleep/wakefulness. *Occupational Medicine, 53,* 89–94.

Aldrin, B. (2009). *Magnificent desolation.* London: Bloomsbury.

Alling, A., Nelson, M., and Silverstone, S. (1993). *Life under glass: The inside story of Biosphere 2.* Oracle, AZ: The Biosphere Press.

Allison, S.T. and Goethals, G.R. (2010). *Heroes: What they do and why we need them.* New York: Oxford University Press.

Altman, I. (1975). *The environment and social behavior.* Monterey: Brooks/Cole.

Altman, I. (1976). Privacy: A conceptual analysis. *Environment and Behavior, 8,* 7–29.

Amiot, C., Vallerand, R.J., and Blanchard, C. (2006). Passion and psychological adjustment: A test of the person-environment fit hypothesis. *Personality and Social Psychology Bulletin, 32,* 220–9.

Amundsen R. (1913). *The South Pole, vol. 1.* (Trans. A.G. Chater). London: John Murray (originally published in Norwegian in 1912).

Andrée, S.A., Strindberg, N., and Fraenkel, K. (1930). *Andrée's story: The complete record of his polar flight, 1897.* (Trans. E. Adams-Ray.) New York: The Viking Press.

Antonenko, D. et al. (2013). Napping to renew learning capacity: Enhanced encoding after stimulation of sleep slow oscillations. *European Journal of Neuroscience, 37,* 1142–51.

Apkarian, A.V. et al. (2005). Human brain mechanisms of pain perception and regulation in health and disease. *European Journal of Pain, 9,* 463–84.

Ardelt, M. (2003). Empirical assessment of a three-dimensional wisdom scale. *Research on Aging, 25,* 275–324.

Arnesen, L., Bancroft, A., and Dahle, C. (2004). *No horizon is so far: Two women and their historic journey across Antarctica.* New York: Penguin Books.

Arnett, J. (1992). Reckless behavior in adolescence: A developmental perspective. *Developmental Review, 12,* 339–73.

Arnett, J. (1995). The young and the reckless: Adolescent reckless behavior. *Current Directions in Psychological Science, 4,* 67–71.

Arnsten, A., Mazure, C.M., and Sinha, R. (2012). This is your brain in meltdown. *Scientific American*, *306* (4), 48–53.

Arrigo, B.A. and Bullock, J.L. (2008). The psychological effects of solitary confinement on prisoners in supermax units: Reviewing what we know and recommending what should change. *International Journal of Offender Therapy and Comparative Criminology*, *52*, 622–40.

Asaoka, S. et al. (2012). The effects of a night-time nap on the error-monitoring functions during extended wakefulness. *Sleep*, *35*, 871–8.

Aschbacher, K. et al. (2013). Good stress, bad stress and oxidative stress: Insights from anticipatory cortisol reactivity. *Psychoneuroendocrinology*, *38*, 1698–708.

Ashcroft, F.M. (2001). *Life at the extremes: The science of survival*. London: Flamingo.

Askew, C. and Field, A.P. (2008). The vicarious learning pathway to fear 40 years on. *Clinical Psychology Review*, *28*, 1249–65.

Atlis, M.M. et al. (2004). Decision processes and interactions during a two-woman traverse of Antarctica. *Environment and Behavior*, *36*, 402–23.

Baird, B., Smallwood, J., and Schooler, J.W. (2011). Back to the future: Autobiographical planning and the functionality of mind-wandering. *Consciousness and Cognition*, *20*, 1604–11.

Baltes, P.B. and Staudinger, U.M. (2000). Wisdom: A metaheuristic (pragmatic) to orchestrate mind and virtue toward excellence. *American Psychologist*, *55*, 122–36.

Baranski, J.V. et al. (2007). Effects of sleep loss on team decision making: Motivational loss or motivational gain? *Human Factors*, *49*, 646–60.

Barger, L.K. et al. (2014). Sleep and cognitive function of crewmembers and mission controllers working 24-h shifts during a simulated 105-day spaceflight mission. *Acta Astronautica*, *93*, 230–42.

Barlow, M., Woodman, T., and Hardy, L. (2013). Great expectations: Different high-risk activities satisfy different motives. *Journal of Personality and Social Psychology*, *105*, 458.

Barnes, C.M. and Hollenbeck, J.R. (2009). Sleep deprivation and decision-making teams: Burning the midnight oil or playing with fire? *Academy of Management Review*, *34*, 56–66.

Barnes, L.L., Harp, D., and Jung, W.S. (2002). Reliability generalization of scores on the Spielberger state-trait anxiety inventory. *Educational and Psychological Measurement*, *62*, 603–18.

Baron, J. (2000). *Thinking and deciding* (3rd edition). Cambridge: Cambridge University Press.

Bartone, P.T. and Adler, A.B. (1999). Cohesion over time in a peacekeeping medical task force. *Military Psychology*, *11*, 85–107.

Bartone, P.T. (1999). Hardiness protects against war-related stress in army reserve forces. *Consulting Psychology Journal: Practice and Research*, *51*, 72–82.

Bates, D.G. and Plog, F. (1976). *Cultural anthropology* (3rd edition). New York: McGraw-Hill.

Bateson, P. and Martin, P. (1999). *Design for a life: How behaviour develops*. London: Jonathan Cape.

Bateson, P. and Martin, P. (2013). *Play, playfulness, creativity and innovation*. Cambridge: Cambridge University Press.

Baumeister, R.F. and Leary, M.R. (1995). The need to belong: Desire for interpersonal attachments as a fundamental human motivation. *Psychological Bulletin*, *117*, 497–529.

Baumeister, R.F., Twenge, J.M., and Nuss, C.K. (2002). Effects of social exclusion on cognitive processes: Anticipated aloneness reduces intelligent thought. *Journal of Personality and Social Psychology*, *83*, 817–27.

Baumeister, R.F. et al. (1998). Self-control depletion: Is the active self a limited resource? *Journal of Personality and Social Psychology*, *74*, 1252–65.

Baumeister, R.F. et al. (2006). Social exclusion impairs self-regulation. *Journal of Personality and Social Psychology, 88,* 589–604.

Beal, D.J. et al. (2003). Cohesion and performance in groups: A meta-analytic clarification of construct relations. *Journal of Applied Psychology, 88,* 989–1004.

Beecher, H.K. (1946). Pain in men wounded in battle. *Annals of Surgery, 123,* 96–105.

Bélanger, J.J. et al. (2013). Driven by fear: The effect of success and failure information on passionate individuals' performance. *Journal of Personality and Social Psychology, 104,* 180–95.

Bénabou, R. and Tirole, J. (2004). Willpower and personal rules. *Journal of Political Economy, 112,* 848–86.

Bennet, G. (1973). Medical and psychological problems in the 1972 singlehanded transatlantic yacht race. *The Lancet, 302,* 747–54.

Bennet, G. (1974). Psychological breakdown at sea: Hazards of singlehanded ocean sailing. *British Journal of Medical Psychology, 47,* 189–210.

Bennet, G. (1983). *Beyond endurance: Survival at the extremes.* London: Martin, Secker and Warburg Ltd.

Bennis, W. (2007). The challenges of leadership in the modern world. *American Psychologist, 62,* 2–5.

Bentall, R.P. (1990). The illusion of reality: A review and integration of psychological research on hallucinations. *Psychological Bulletin, 107,* 82–95.

Ben-Zur, H. and Zeidner, M. (2009). Threat to life and risk-taking behaviors: A review of empirical findings and explanatory models. *Personality and Social Psychology Review, 13,* 109–28.

Berto, R. (2005). Exposure to restorative environments helps restore attentional capacity. *Journal of Environmental Psychology, 25,* 249–59.

Bexton, W.H., Heron, W., and Scott, T.H. (1954). Effects of decreased variation in the sensory environment. *Canadian Journal of Psychology, 8,* 70–6.

Bhargava, R., Mukerji, S., and Sachdeva, U. (2000). Psychological impact of the Antarctic winter on Indian expeditioners. *Environment and Behavior, 32,* 111–27.

Bhasin, M.K. et al. (2013). Relaxation response induces temporal transcriptome changes in energy metabolism, insulin secretion and inflammatory pathways. *PLoS ONE, 8,* e62817.

Binder, E.B. et al. (2008). Association of *FKBP5* polymorphisms and childhood abuse with risk of posttraumatic stress disorder symptoms in adults. *JAMA, 299,* 1291–305.

Bishop, S.L., Morphew, M.E., and Kring, J.P. (2000). Avoiding risky teams in risky environments. *Proceedings of the Human Factors and Ergonomics Society Annual Meeting, 44,* 800–3.

Bjorvatn, B., Kecklund, G., and Åkerstedt, T. (1998). Rapid adaptation to night work at an oil platform, but slow readaptation following return home. *Journal of Occupational and Environmental Medicine, 40,* 601–8.

Blaivas, A.J. et al. (2007). Quantifying microsleep to help assess subjective sleepiness. *Sleep Medicine, 8,* 156–9.

Blanke, O.F., Arzyi, S., and Landis, T. (2008). Illusory perceptions of the human body and self. In G. Goldenberg and B. Miller (Eds) *Handbook of clinical neurology, neuropsychology and behavioral neurology, vol. 88* (pp. 429–58). Paris: Elsevier.

Blaszczynski, A., McConaghy, N., and Frankova, A. (1990). Boredom proneness in pathological gambling. *Psychological Reports, 67,* 35–42.

Blum, A. (2005). *Breaking trail.* New York: Scribner.

Boggild, H. and Knutsson, A. (1999). Shift work, risk factors and cardiovascular disease. *Scandinavian Journal of Work and Environmental Health, 25,* 85–99.

Bonanno, G.A. (2004). Loss, trauma, and human resilience. *American Psychologist, 59*, 20–8.

Bonanno, G.A. (2005). Resilience in the face of potential trauma. *Current Directions in Psychological Science, 14*, 135–8.

Bonanno, G.A. et al. (2006). Psychological resilience after disaster: New York City in the aftermath of the September 11th terrorist attack. *Psychological Science, 17*, 181–6.

Bond, M.R. and Simpson, K.H. (2006). *Pain: Its nature and treatment*. Edinburgh: Churchill Livingstone.

Bood, S.Å. et al. (2006). Eliciting the relaxation response with the help of flotation-rest (restricted environmental stimulation technique) in patients with stress-related ailments. *International Journal of Stress Management, 13*, 154–75.

Booth, J.N., Koren, S.A., and Persinger, M.A. (2005). Increased feelings of the sensed presence and increased geomagnetic activity at the time of the experience during exposures to transcerebral weak complex magnetic fields. *International Journal of Neuroscience, 115*, 1053–79.

Boss, P.G. (1999). *Ambiguous loss: Learning to live with unresolved grief*. Cambridge, MA: Harvard University Press.

Boss, P.G. (2002). Ambiguous loss: Working with families of the missing. *Family Processes, 41*, 14–17.

Boston, P.J. (2004). Exploring the deep underground. In S.J. Dick and K.L. Cowing (Eds) *Risk and exploration: Earth, sea, and the stars. NASA Administrator's Symposium September 26–29, 2004, Naval Postgraduate School Monterey, California* (pp. 55–60). Washington, DC: NASA.

Bowley, G. (2010). *No way down: Life and death on K2*. London: Penguin.

Bown, S. (2012). *The last Viking: The life of Roald Amundsen, conqueror of the South Pole*. London: Aurum Press.

Breashears, D. (2000). *High exposure: An enduring passion for Everest and unforgiving places*. New York: Simon and Schuster.

Brefczynski-Lewis, J.A. et al. (2007). Neural correlates of attentional expertise in long-term meditation practitioners. *Proceedings of the National Academy of Sciences USA, 104*, 11483–8.

Bright, J.E.H. et al. (2009). Chance events in career development: Influence, control and multiplicity. *Journal of Vocational Behavior, 75*, 14–25.

Britton, A. and Shipley, M.J. (2010). Bored to death? *International Journal of Epidemiology, 39*, 370–1.

Broadbent, N.D. (2007). From ballooning in the arctic to 10,000-foot runways in Antarctica: Lessons from historic archaeology. In I. Krupnik, M.A. Lang, and S.E. Miller (Eds) *Smithsonian at the Poles* (pp. 49–60). Washington, DC: Smithsonian Institution Scholarly Press.

Brugger, P. (2006). From phantom limb to phantom body: Varieties of extracorporeal awareness. In G. Knoblich et al. (Eds) *Human body perception from the inside out* (pp. 171–209). Oxford: Oxford University Press.

Brugger, P., Regard, M., Landis, T., and Oelz, O. (1999). Hallucinatory experiences in extreme-altitude climbers. *Neuropsychiatry, Neuropsychology, and Behavioral Neurology, 12*, 67–71.

Brugman, G.M. (2006). *Wisdom and aging*. Amsterdam: Elsevier.

Brymer, E. and Gray, T. (2010). Developing an intimate 'relationship' with nature through extreme sports participation. *Leisure/Loisir, 34*, 361–74.

Brymer, E. and Oades L.G. (2009). Extreme sports: A positive transformation in courage and humility. *Journal of Humanistic Psychology, 49*, 114–26.

Brymer, E. and Schweitzer, R. (2013). Extreme sports are good for your health: A phenomenological understanding of fear and anxiety in extreme sport. *Journal of Health Psychology, 18*, 477–87.

Buchanan, R. (2000). A natural death. *Outside Magazine*, 1 June.

Bucks, R.S. et al. (2008). Selective effects of upper respiratory tract infection on cognition, mood and emotion processing: A prospective study. *Brain, Behavior, and Immunity*, 22, 399–407.

Buehler, R., Griffin, D., and Ross, M. (1994). Exploring the 'planning fallacy': Why people underestimate their task completion times. *Journal of Personality and Social Psychology*, 67, 366–81.

Buguet, A. (2007). Sleep under extreme environments: Effects of heat and cold exposure, altitude, hyperbaric pressure and microgravity in space. *Journal of the Neurological Sciences*, 262, 145–52.

Burger, J.M. (1995). Individual differences in preference for solitude. *Journal of Research in Personality*, 29, 85–108.

Burgess, E.O. et al. (2001). Surfing for sex: Studying involuntary celibacy using the internet. *Sexuality and Culture*, 5, 5–30.

Burke C.S. et al. (2006). Understanding team adaptation: A conceptual analysis and model. *Journal of Applied Psychology*, 91, 1189–207.

Burke, C.S. et al. (2008). Stress and teams: How stress affects decision making at the team level. In P.A. Hancock and J.L. Szalma (Eds) *Performance under stress* (pp. 181–208). Aldershot: Ashgate Publishing.

Burns, R. (2001). *Just tell them I survived! Women in Antarctica*. Crows Nest: Allen and Unwin.

Burroughs, B. (1999). *Dragonfly and the crisis aboard Mir*. London: Fourth Estate.

Burtscher, M. (2012). Climbing the Himalayas more safely. *British Medical Journal*, 344, e3778.

Bushnell, M.C., Ceko, M., and Low, L.A. (2013). Cognitive and emotional control of pain and its disruption in chronic pain. *Nature Reviews Neuroscience*, 14, 502–11.

Byrd, R.E. (1938). *Alone*. New York: G.P. Putnam's Sons.

Byrd, R.E. and Gould, L.M. (1930). *Little America, aerial exploration in the Antarctic: The flight to the South Pole*. New York: G.P. Putnam's Sons.

Cabrol, N. (2004). Exploring the world's highest lakes. In S.J. Dick and K.L. Cowing (Eds) *Risk and exploration: Earth, sea, and the stars. NASA Administrator's Symposium September 26–29, 2004, Naval Postgraduate School Monterey, California* (pp. 61–7). Washington, DC: NASA.

Cacioppo, J.T. and Patrick, W. (2008). *Loneliness: Human nature and the need for social connection*. New York: W.W. Norton.

Cahn, B.R. and Polich, J. (2006). Meditation states and traits: EEG, ERP, and neuroimaging studies. *Psychological Bulletin*, 132, 180–211.

Cain, S. (2012). *Quiet: The power of introverts in a world that can't stop talking*. New York: Crown.

Cannon-Bowers, J.A. and Salas, E. (1998). Team performance and training in complex environments: Recent findings from applied research. *Current Directions in Psychological Science*, 7, 83–7.

Cappuccio, F.P. et al. (2011). Sleep duration predicts cardiovascular outcomes: A systematic review and meta-analysis of prospective studies. *European Heart Journal*, 32, 1484–92.

Carlson, L.E. et al. (2004). Mindfulness-based stress reduction in relation to quality of life, mood, symptoms of stress and levels of cortisol, dehydroepiandrosterone sulfate (DHEAS) and melatonin in breast and cancer outpatients. *Psychoneuroendocrinology*, 29, 448–74.

Carlson, S.R., Johnson, S.C., and Jacobs, P.C. (2010). Disinhibited characteristics and binge drinking among university student drinkers. *Addictive Behaviors*, 35, 242–51.

Carriere, J.S.A., Cheyne, J.A., and Smilek, D. (2008). Everyday attention lapses and memory failures: The affective consequences of mindlessness. *Consciousness and Cognition*, 17, 835–47.

Caruso, C.C., Lusk, S.L., and Gillespie, B.W. (2004). Relationship of work schedules to gastrointestinal diagnoses, symptoms, and medication use in auto factory workers. *American Journal of Industrial Medicine*, 46, 586–98.

Castanier, C., Le Scanff, C., and Woodman, T. (2011). Mountaineering as affect regulation: The moderating role of self-regulation strategies. *Anxiety, Stress, and Coping*, 24, 75–89.

Casteret, N. (1938). *Ten years under the earth*. New York: The Greystone Press.

Cazenave, N., Le Scanff, C., and Woodman, T. (2007). Psychological profiles and emotional regulation characteristics of women engaged in risk-taking sports. *Anxiety, Stress, and Coping*, 20, 421–35.

Ceja, L. and Navarro, J. (2011). Dynamic patterns of flow in the workplace: Characterising within-individual variability using a complexity science approach. *Journal of Organizational Behavior*, 32, 627–51.

Cernan, E. and Davis, D.A. (1999). *The last man on the Moon*. New York: St Martins.

Cervero, F. (2012). *Understanding pain*. Cambridge, MA: MIT Press.

Chaikin, A. (1994). *A man on the Moon: The triumphant story of the Apollo space program*. New York: Viking.

Chan, D. and Rossor, M.N. (2002). '—but who is that on the other side of you?' Extracampine hallucinations revisited. *The Lancet*, 360, 2064–6.

Chen, C-F. and Chen, C-W. (2011). Speeding for fun? Exploring the speeding behavior of riders of heavy motorcycles using the theory of planned behavior and psychological flow theory. *Accident Analysis and Prevention*, 43, 983–90.

Cherry-Garrard, A. (2010). *The worst journey in the world*. London: Vintage (originally published 1922).

Cheyne, J.A. (2009). Sensed presences in extreme contexts. *Skeptic*, 15, 68–71.

Cheyne, J.A. (2012). Sensed presences. In J.D. Blom and I.E.C. Sommer (Eds) *Hallucinations: Research and practice* (pp. 219–34). New York: Springer.

Chi, M.T.H., Glaser, R., and Farr, M.J. (Eds) (1988). *The nature of expertise*. Hillsdale: Erlbaum.

Churton, T. (2011). *Aleister Crowley: Spiritual revolutionary, romantic explorer, occult master and spy*. London: Watkins Publishing.

Cialdini, R.B. and Goldstein, N.J. (2004). Social influence: Compliance and conformity. *Annual Review of Psychology*, 55, 591–621.

Clark, B. and Graybiel, A. (1957). The break-off phenomenon: A feeling of separation from the earth experienced by pilots at high altitude. *The Journal of Aviation Medicine*, 28, 121.

Clearwater, Y.A. and Coss, R.G. (1991). Functional esthetics to enhance well-being in isolated and confined settings. In A.A. Harrison, Y.A. Clearwater, and C.P. McKay (Eds) *From Antarctica to outer space* (pp. 331–48). New York: Springer-Verlag.

Coffey, M. (2004). *Where the mountain casts its shadow: The personal costs of climbing*. London: Random House.

Cohen, D.J. et al. (2013). Quantitative electroencephalography during rapid eye movement (REM) and non-REM sleep in combat-exposed veterans with and without post-traumatic stress disorder. *Journal of Sleep Research*, 22, 76–82.

Cohen M.S., Freeman, J.T., and Thompson, B.T. (1997). Training the naturalistic decision maker. In C.E. Zsambok and G. Klein (Eds) *Naturalistic decision making* (pp. 257–68). Mahwah, NJ: Erlbaum.

Colasanti, A. et al. (2008). Carbon dioxide-induced emotion and respiratory symptoms in healthy volunteers. *Neuropsychopharmacology*, 33, 3103–10.

Collyer, S.C. and Malecki, G.S. (1998). Tactical decision making under stress: History and overview. In J.A. Cannon-Bowers and E. Salas (Eds) *Making decisions under stress: Implications for individual and team training* (pp. 3–15). Washington, DC: American Psychological Association.

Colman, R.J. et al. (2009). Caloric restriction delays disease onset and mortality in rhesus monkeys. *Science, 325,* 201–4.

Colvard, D.F. and Colvard, L.Y. (2003). A study of panic in recreational scuba divers. *The Undersea Journal, First Quarter,* 40–4.

Conefrey, M. (2011). *How to climb Mont Blanc in a skirt.* London: OneWorld.

Connor, J. et al. (2001). The role of driver sleepiness in car crashes: A systematic review of epidemiological studies. *Accident Analysis* and *Prevention, 33,* 31–41.

Connor, K.M. and Davidson, J.R.T. (2003). Development of a new resilience scale: The Connor–Davidson Resilience Scale (CD-RISC). *Depression and Anxiety, 18,* 76–82.

Cook, F.A. (1900). *Through the first Antarctic night.* New York: Doubleday and McClure Co.

Coplan, R.J. and Weeks, M. (2010a). Unsociability in the preference for solitude in childhood. In K.H. Rubin and R.J. Coplan (Eds) *The development of shyness and social withdrawal* (pp. 64–83). New York: Guilford Press.

Coplan, R.J. and Weeks, M. (2010b). Unsociability in middle childhood: Conceptualization, assessment, and associations with socioemotional functioning. *Merrill-Palmer Quarterly, 56,* 105–30.

Coplan, R.J. et al. (2004). Do you 'want' to play? Distinguishing between conflicted shyness and social disinterest in early childhood. *Developmental Psychology, 40,* 244–58.

Corbett, S. (2004). Yes, it is a lovely morning. Now why don't you just go to Hell. *Outside Magazine,* 2 May.

Courtauld, A. (1932). Living alone under polar conditions. *Polar Record, 1,* 66–74.

Cramer, K.M. and Lake, R.P. (1998). The preference for solitude scale: Psychometric properties and factor structure. *Comparative and General Pharmacology, 24,* 193–9.

Cravalho, M.A. (1996). Toast on ice: The ethnopsychology of the winter-over experience in Antarctica. *Ethos, 24,* 628–56.

Critcher, C.R. and Gilovich, T. (2010). Inferring attitudes from mindwandering. *Personality and Social Psychology Bulletin, 36,* 1255–66.

Cronin, C. (1991). Sensation seeking among mountain climbers. *Personality and Individual Differences, 12,* 653–4.

Cross, P. (1977). Not can but will college teaching be improved. *New Directions for Higher Education, 17,* 1–15.

Crust, L. (2008). A review and conceptual re-examination of mental toughness: implications for future researchers. *Personality and Individual Differences, 45,* 576–83.

Crust, L. and Clough, P.J. (2005). Relationship between mental toughness and physical endurance. *Perceptual and Motor Skills, 100,* 192–4.

Crust, L. and Swann, C. (2013). The relationship between mental toughness and dispositional flow. *European Journal of Sport Science, 13,* 215–20.

Csikszentmihalyi, M. (1990). *Flow: The psychology of optimal experience.* New York: Harper and Row.

Culp, N.A. (2006). The relations of two facets of boredom proneness with the major dimensions of personality. *Personality and Individual Differences, 41,* 999–1007.

Curtis, V. and Biran, A. (2001). Dirt, disgust, and disease. *Perspectives in Biology and Medicine, 44,* 17–31.

Dahlen, E.R. et al. (2005). Driving anger, sensation seeking, impulsiveness, and boredom proneness in the prediction of unsafe driving. *Accident Analysis and Prevention*, 37, 341–8.

Dahlen, E.R. and White, R.P. (2006). The Big Five factors, sensation seeking, and driving anger in the prediction of unsafe driving. *Personality and Individual Differences*, 41, 903–15.

Damrad-Frye, R. and Laird, J.D. (1989). The experience of boredom: The role of self-perception of attention. *Journal of Personality and Social Psychology*, 57, 315–20.

Darlington, J. (1957). *My Antarctic honeymoon*. London: Frederick Muller.

Darwin, C. (1958). *The autobiography of Charles Darwin and selected letters*. (Ed. F. Darwin.) New York: Dover.

Davidson, R.J. et al. (2003). Alterations in brain and immune function produced by mindfulness meditation. *Psychosomatic Medicine*, 65, 564–70.

Davis, M. et al. (2010). Phasic vs sustained fear in rats and humans: Role of the extended amygdala in fear vs anxiety. *Neuropsychopharmacology Reviews*, 35, 105–35.

Davis, W. (2012). *Into the silence: The Great War, Mallory and the conquest of Everest*. New York: Alfred A. Knopf.

Dawson, D. and Reid, K. (1997). Fatigue, alcohol and performance impairment. *Nature*, 388, 235.

De Jong, B.A. and Dirks, K.T. (2012). Beyond shared perceptions of trust and monitoring in teams: Implications of asymmetry and dissensus. *Journal of Applied Psychology*, 97, 391–406.

Décamps, G. and Rosnet, E. (2005). A longitudinal assessment of psychological adaptation during a winter-over in Antarctica. *Environment and Behavior*, 37, 418–35.

DeChurch, L.A. and Mesmer-Magnus, J.R. (2010). The cognitive underpinnings of effective teamwork: A meta-analysis. *Journal of Applied Psychology*, 95, 32–53.

Delle Fave, A., Bassi, M., and Massimini, F. (2003). Quality of experience and risk perception in high-altitude rock climbing. *Journal of Applied Sport Psychology*, 15, 82–98.

Dement, W.C. and Vaughan, C. (1999). *The promise of sleep*. New York: Delacorte.

Dias, C., Cruz, J.F., and Fonseca, A.M. (2012). The relationship between multidimensional competitive anxiety, cognitive threat appraisal, and coping strategies: A multi-sport study. *International Journal of Sport and Exercise Psychology*, 10, 52–65.

Dickerson, S.S. and Kemeny, M.E. (2004). Acute stressors and cortisol responses: A theoretical integration and synthesis of laboratory research. *Psychological Bulletin*, 130, 355–91.

Diehm, R. and Armatas, C. (2004). Surfing: An avenue for socially acceptable risk-taking, satisfying needs for sensation seeking and experience seeking. *Personality and Individual Differences*, 36, 663–77.

Dijk, D.J. et al. (2001). Sleep, performance, circadian rhythms, and light-dark cycles during two space shuttle flights. *American Journal of Physiology-Regulatory, Integrative and Comparative Physiology*, 281, R1647–64.

Dion, K.L. (2004). Interpersonal and group processes in long-term spaceflight crews: Perspectives from social and organizational psychology. *Aviation, Space, and Environmental Medicine*, 75 (Supplement 1), C36–43.

Dirks, K.T. (2000). Trust in leadership and team performance: Evidence from NCAA basketball. *Journal of Applied Psychology*, 85, 1004–12.

Dirks, K.T. and Ferrin, D.L. (2002). Trust in leadership: Meta-analytic findings and implications for research and practice. *Journal of Applied Psychology*, 87, 611–28.

Donnelly, D. et al. (2001). Involuntary celibacy: A life course analysis. *Journal of Sex Research*, 38, 159–69.

Downey, G. et al. (1998). The self-fulfilling prophecy in close relationships: Rejection sensitivity and rejection by romantic partners. *Journal of Personality and Social Psychology*, 75, 545–60.

Dreyfus, H.L. (1997). Intuitive, deliberative, and calculative models of expert performance. In C. Zsambok and G. Klein (Eds) *Naturalistic decision making* (pp. 17–28). Mahwah, NJ: Lawrence Erlbaum Associates Inc.

Driskell, J.E. et al. (2006). What makes a good team player? Personality and team effectiveness. *Group Dynamics: Theory, Research, and Practice*, 10, 249–71.

Drummond, S.P.A., Paulus, M.P., and Tapert, S.F. (2006). Effects of two nights sleep deprivation and two nights recovery sleep on response inhibition. *Journal of Sleep Research*, 15, 261–5.

Duangpatra, K.N.K., Bradley, G.L., and Glendon, A.I. (2009). Variables affecting emerging adults' self-reported risk and reckless behaviors. *Journal of Applied Developmental Psychology*, 30, 298–309.

Duckworth, A.L. et al. (2007). Grit: Perseverance and passion for long-term goals. *Journal of Personality and Social Psychology*, 92, 1087–101.

Dumas, L.J. (2011). When costs approach infinity: Microeconomic theory, security, and dangerous technologies. In M. Chatterji, C. Bo, and R. Misra (Eds) *Frontiers of peace economics and peace science (contributions to conflict management, peace economics and development, volume 16)* (pp. 59–71). Bingley: Emerald Group Publishing Ltd.

Dunning, D. et al. (2003). Why people fail to recognize their own incompetence. *Current Directions in Psychological Science*, 12, 83–7.

Durmer, J.S. and Dinges, D.F. (2005). Neurocognitive consequences of sleep deprivation. *Seminars in Neurology*, 25, 117–29.

Dusek, J.A. et al. (2008). Genomic counter-stress changes induced by the relaxation response. *PLoS ONE*, 3, e2576.

Dutton, D.G. and Aron, A.P. (1974). Some evidence for heightened sexual attraction under conditions of high anxiety. *Journal of Personality and Social Psychology*, 30, 510–17.

Eastwood, J.D. et al. (2012). The unengaged mind: Defining boredom in terms of attention. *Perspectives on Psychological Science*, 7, 482–95.

Ebaugh, H.E.F. (1988). *Becoming an ex: The process of role exit*. Chicago: University of Chicago Press.

Egan, S. and Stelmack, R.M. (2003). A personality profile of Everest climbers. *Personality and Individual Differences*, 34, 1491–4.

Einarsen, S., Aasland, M.S., and Skogstad, A. (2007). Destructive leadership behaviour: A definition and conceptual model. *The Leadership Quarterly*, 18, 207–16.

Eisenberger, N.I., Lieberman, M.D., and Williams, K.D. (2003). Does rejection hurt? An fMRI study of social exclusion. *Science*, 302, 290–2.

Elder, G.H. and Clipp, E.C. (1989). Combat experience and emotional health: Impairment and resilience in later life. *Journal of Personality*, 57, 311–41.

Ellemers, N., Spears, R., and Doosje, B. (2002). Self and social identity. *Annual Review of Psychology*, 53, 161–86.

Elwood, L.S. and Olatunji, B.O. (2009). A cross-cultural perspective on disgust. In B.O. Olatunji and D. McKay, (Eds) *Disgust and its disorders: Theory, assessment, and treatment implications* (pp. 99–122). Washington, DC: American Psychological Association.

Endsley, M.R. (1995). Toward a theory of situation awareness in dynamic systems. *Human Factors: The Journal of the Human Factors and Ergonomics Society*, 37, 32–64.

Endsley, M.R. (2000). Theoretical underpinnings of situation awareness: A critical review. In M.R. Endsley and D.J. Garland (Eds) *Situation awareness: Analysis and measurement* (pp. 3–33). Mahwah, NJ: Lawrence Erlbaum Associates Inc.

Ericsson, K.A. (2006). The influence of experience and deliberate practice on the development of superior expert performance. *The Cambridge handbook of expertise and expert performance* (pp. 683–703). Cambridge: Cambridge University Press.

Ericsson, K.A. and Charness, N. (1994). Expert performance: Its structure and acquisition. *American Psychologist, 49*, 725–47.

Ericsson, K.A. and Lehman, A.C. (1996). Expert and exceptional performance: Evidence of maximal adaptation to task constraints. *Annual Review of Psychology, 47*, 273–305.

Evans, G.W., Lepore, S.J., and Allen, K.M. (2000). Cross-cultural differences in tolerance for crowding: Fact or fiction? *Journal of Personality and Social Psychology, 79*, 204.

Everett, C.C. (1891). *Ethics for young people*. Boston, MA: Ginn and Company.

Ewert, A.W. (1994). Playing the edge: Motivation and risk taking in a high-altitude wilderness-like environment. *Environment and Behavior, 26*, 3–24.

Faraut, B. et al. (2011). Benefits of napping and an extended duration of recovery sleep on alertness and immune cells after acute sleep restriction. *Brain, Behavior, and Immunity, 25*, 16–24.

Fear, N.T. et al. (2010). What are the consequences of deployment to Iraq and Afghanistan on the mental health of the UK armed forces? A cohort study. *The Lancet, 375*, 1783–97.

Feder, A., Nestler, E.J., and Charney, D.S. (2009). Psychobiology and molecular genetics of resilience. *Nature Reviews Neuroscience, 10*, 446–57.

Ferguson, M.J. (1970). *Use of the Ben Franklin submersible as a space station analog. Vol. II: Psychology and physiology*. Bethpage, NY: Grumman.

Ferrell, J. (2004). Boredom, crime, and criminology. *Theoretical Criminology, 8*, 287–302.

Fiennes, R. (2007). *Mad, bad and dangerous to know*. London: Hodder and Stoughton.

Fiennes, R. (2013). *Cold: Extreme adventures at the lowest temperatures on Earth*. London: Simon and Schuster.

Finch, P. (2008). *Raising the dead*. London: HarperSport.

Finney, B. (1991). Scientists and seamen. In A.A. Harrison, Y.A. Clearwater, and C.P. McKay (Eds) *From Antarctica to outer space* (pp. 89–102). New York: Springer-Verlag.

Fisher, C.D. (1993). Boredom at work: A neglected concept. *Human Relations, 46*, 395–417.

Fisher, C.D. (1998). Effects of external and internal interruptions on boredom at work: Two studies. *Journal of Organizational Behavior, 19*, 503–22.

Fitch, J.L. and Ravlin, E.C. (2005). Willpower and perceived behavioral control: Influences on the intention–behavior relationship and postbehavior attributions. *Social Behavior and Personality, 33*, 105–24.

Flanagan, J.C. (1954). The critical incident technique. *Psychological Bulletin, 51*, 327–58.

Fleming, F. (2000). *Ninety degrees north: The quest for the North Pole*. London: Granta.

Flyvbjerg, B., Garbuio, M., and Lovallo, D. (2009). Delusion and deception in large infrastructure projects: Two models for explaining and preventing executive disaster. *California Management Review, 51*, 170–93.

Franco, Z.E., Blau, K., and Zimbardo, P.G. (2011). Heroism: A conceptual analysis and differentiation between heroic action and altruism. *Review of General Psychology, 15*, 99–113.

Frankl, V. (1984). *Man's search for meaning*. New York: Simon and Schuster (originally published 1946).

Franklin, J. (2011). *The 33: The ultimate account of the Chilean miners' dramatic rescue*. London: Transworld Publishers.

Fraser, T.M. (1968). *The intangibles of habitability during long duration space missions*. Washington, DC: NASA.

Fredrickson, B.L. et al. (2003). What good are positive emotions in crises? A prospective study of resilience and emotions following the terrorist attacks on the United States on September 11th, 2001. *Journal of Personality and Social Psychology, 84*, 365–76.

Frost, O.W. (1995). Vitus Bering and Georg Steller: Their tragic conflict during the American expedition. *Pacific Northwest Quarterly, 86*, 3–16.

Frost, O.W. (2003). *Bering: The Russian discovery of America*. New Haven, CT: Yale University Press.

Gailliot, M.T. et al. (2007). Self-control relies on glucose as a limited energy source: Willpower is more than a metaphor. *Journal of Personality and Social Psychology, 92*, 325–36.

Galea, S. et al. (2003). Trends of probably post-traumatic stress disorder in New York City after the September 11th terrorist attacks. *American Journal of Epidemiology, 158*, 514–24.

Gard, T. et al. (2012). Pain attenuation through mindfulness is associated with decreased cognitive control and increased sensory processing in the brain. *Cerebral Cortex, 22*, 2692–702.

Gecas, V. (1989). The social psychology of self-efficacy. *Annual Review of Sociology, 15*, 291–316.

Geiger, J. (2009). *The third man factor*. Edinburgh: Canongate.

Gilbert, J.A. et al. (2012). Toxic versus cooperative behaviors at work: The role of organizational culture and leadership in creating community-centred organizations. *International Journal of Leadership Studies, 7*, 29–47.

Glouberman, S. (2009). Knowledge transfer and the complex story of scurvy. *Journal of Evaluation in Clinical Practice, 15*, 553–7.

Goldberg, Y.K. et al. (2011). Boredom: An emotional experience distinct from apathy, anhedonia, or depression. *Journal of Social and Clinical Psychology, 30*, 647–66.

Gonsalkorale, K. and Williams, K.D. (2007). The KKK won't let me play: Ostracism even by a despised outgroup hurts. *European Journal of Social Psychology, 37*, 1176–85.

Gooderham, P. (2009). The distinction between gross negligence and recklessness in English criminal law. *Journal of the Royal Society of Medicine, 102*, 358.

Gouin, J.P., Hantsoo, L., and Kiecolt-Glaser, J.K. (2008). Immune dysregulation and chronic stress among older adults: A review. *Neuroimmunomodulation, 15*, 251–9.

Grandner, M.A. et al. (2010). Mortality associated with short sleep duration: The evidence, the possible mechanisms, and the future. *Sleep Medicine Reviews, 14*, 191–203.

Grassian, S. (1983). Psychopathological effects of solitary confinement. *American Journal of Psychiatry, 140*, 1450–4.

Greely, A.W. (1886). *Three years of Arctic service: An account of the Lady Franklin Bay expedition of 1881–1884 and the attainment of farthest North, Vol 1*. New York, NY: Charles Scribner's Sons.

Green, J.L., Sawaya, F.J., and Dollar, A.L. (2011). The effects of caloric restriction on health and longevity. *Current Treatment Options in Cardiovascular Medicine, 13*, 326–34.

Griffiths, T. (2007). *Slicing the silence: Voyaging to Antarctica*. Sydney, Australia: University of New South Wales Press.

Grossman, P. et al. (2004). Mindfulness-based stress reduction and health benefits: A meta-analysis. *Journal of Psychosomatic Research, 57*, 35–43.

Gruter, M. and Masters, R.D. (1986). Ostracism: A social and biological phenomenon. *Ethology and Sociobiology, 7*, 149–395.

Gully, S.M., et al. (2002). A meta-analysis of team-efficacy, potency, and performance: Inter-dependence and level of analysis as moderators of observed relationships. *Journal of Applied Psychology*, 87, 819–32.

Guly, H.R. (2013). Use and abuse of alcohol and other drugs during the heroic age of Antarctic exploration. *History of Psychiatry*, 24, 94–105.

Guttridge, L.F. (2000). *Ghosts of Cape Sabine: The harrowing true story of the Greely expedition.* New York: G.P. Putnam and Sons.

Hadfield, C. (2013). *An astronaut's guide to life on earth.* London: Macmillan.

Haggard, E.A. (1973). Some effects of geographic and social isolation in natural settings. In J.E. Rasmussen (Ed.) *Man in isolation and confinement* (pp. 99–144). Chicago: Aldine Publishing Company.

Halagappa, V.K.M. et al. (2007). Intermittent fasting and caloric restriction ameliorate age-related behavioral deficits in the triple-transgenic mouse model of Alzheimer's disease. *Neurobiology of Disease*, 26, 212–20.

Hale, D. (2009). *The final dive: The life and death of 'Buster' Crabb.* Stroud: The History Press.

Hall, B. (2012). *Team spirit: Life and leadership on one of the world's toughest yacht races.* London: Bloomsbury.

Hames, J.L., Hagan, C.R., and Joiner, T.E. (2013). Interpersonal processes in depression. *Annual Review of Clinical Psychology*, 9, 355–77.

Hamilton-Paterson, J. (2010). *Empire of the clouds.* London: Faber and Faber.

Hannah, S.T., Campbell, D.J., and Matthews, M.D. (2010). Advancing a research agenda for leadership in dangerous contexts. *Military Psychology*, 22, S157–89.

Harrer, H. (2005). *The white spider.* London: Harper Perennial (originally published 1959).

Harris, M.B. (2000). Correlates and characteristics of boredom proneness and boredom. *Journal of Applied Social Psychology*, 30, 576–98.

Harrison, A.A. (2001). *Spacefaring: The human dimension.* Berkley: University of California Press.

Harrison, Y. and Horne, J.A. (1999). One night of sleep loss impairs innovative thinking and flexible decision making. *Organizational Behavior and Human Decision Processes*, 78, 128–45.

Harrison, Y. and Horne, J.A. (2000). The impact of sleep deprivation on decision making: A review. *Journal of Experimental Psychology: Applied*, 6, 236–49.

Hawkley, L.C. et al. (2003). Loneliness in everyday life: Cardiovascular activity, psychosocial context, and health behaviors. *Journal of Personality and Social Psychology*, 85, 105–20.

Hayashi, M., Ito, S., and Hori, T. (1999). The effects of a 20-min nap at noon on sleepiness, performance and EEG activity. *International Journal of Psychophysiology*, 32, 173–80.

Heinrich, L.M. and Gullone, E. (2006). The clinical significance of loneliness: A literature review. *Clinical Psychology Review*, 26, 695–718.

Hellwarth, B. (2012). *Sealab: America's forgotten quest to live and work on the ocean floor.* New York: Simon and Schuster.

Helweg-Larsen, M. and Shepperd, J.A. (2001). Do moderators of the optimistic bias affect personal or target risk estimates? A review of the literature. *Personality and Social Psychology Review*, 5, 74–95.

Hersch, M.H. (2012). Space madness: The dreaded disease that never was. *Endeavour*, 36, 32–40.

Hill, K. and Linden, D.E. (2013). Hallucinatory experiences in non-clinical populations. In R. Jardri et al. (Eds) *The neuroscience of hallucinations* (pp. 21–41). New York: Springer.

Hills, P. and Argyle, M. (2001). Happiness, introversion-extraversion and happy introverts. *Personality and Individual Differences*, 30, 595–608.

Hodges, B., Regehr, G., and Martin, D. (2001). Difficulties in recognizing one's own incompetence: Novice physicians who are unskilled and unaware of it. *Academic Medicine*, 76, S87–9.

Hoeksema-van Orden, C.Y.D., Gaillard, A.W.K., and Buunk, B.P. (1998). Social loafing under fatigue. *Journal of Personality and Social Psychology*, 75, 1179–90.

Hofer-Tinguely, G. et al. (2005). Sleep inertia: Performance changes after sleep, rest and active waking. *Cognitive Brain Research*, 22, 323–31.

Hoffman, R. (2001). Human psychological performance in cold environments. In R. Zajtchuk and R.F. Bellamy (Eds) *Textbook of military medicine* (pp. 383–410). Washington, DC: Department of the Army, Office of the Surgeon General, and Borden Institute.

Hogan, R. (2006). *Personality and the fate of organizations*. Hillsdale, NJ: Erlbaum.

Hogg, M.A. (1993). Group cohesiveness: A critical review and some new directions. *European Review of Social Psychology*, 4, 85–111.

Holahan, C.J. et al. (2005). Stress generation, avoidance coping, and depressive symptoms: A 10-year model. *Journal of Consulting and Clinical Psychology*, 73, 658–66.

Hoorens, V. (1993). Self-enhancement and superiority biases in social comparison. *European Review of Social Psychology*, 4, 113–39.

Horsburgh, V.A. et al. (2009). A behavioural genetic study of mental toughness and personality. *Personality and Individual Differences*, 46, 100–5.

Hovland, C.I., Janis, I., and Kelley, H.H. (1953). *Communication and persuasion: Psychological studies of opinion change*. New Haven, CT: Yale University Press.

Hughes, V. (2012). The roots of resilience. *Nature*, 490, 165–7.

Hunter, C. (2011). *Extreme risk: A life fighting the bombmakers*. London: Corgi.

Huntford, R. (1999). *Scott and Amundsen: Their race to the South Pole* (updated edition). London: Abacus.

Hurdiel, R. et al. (2014). Sleep restriction and degraded reaction-time performance in Figaro solo sailing races. *Journal of Sports Sciences*, 32, 172–4.

Hylton, H. (2007). Why astronauts don't like shrinks. *Time Magazine*, 8 February.

Imbernon, E. et al. (1993). Effects on health and social well-being of on-call shifts. *Journal of Occupational Medicine*, 35, 1131–7.

Ingre, M. et al. (2006). Subjective sleepiness and accident risk: Avoiding the ecological fallacy. *Journal of Sleep Research*, 15, 142–8.

Inoue, N., Ichiyo M., and Ohshima, H. (2004). Group interactions in SFINCSS-99: Lessons for improving behavioral support programs. *Aviation, Space, and Environmental Medicine*, 75, C28–35.

Institute of Medicine (2004). *Dietary reference intakes: Water, potassium, sodium, chloride, and sulfate*. Washington, DC: National Academies Press.

Jackson, J.S.H. and Blackman, R. (1994). A driving-simulator test of Wilde's risk homeostasis theory. *Journal of Applied Psychology*, 79, 950–8.

Jackson, S.A. and Csikszentmihalyi, M. (1999). *Flow in sports: The keys to optimal experiences and performances*. Champaign: Human Kinetics.

Jackson S.A. and Eklund, R.C. (2004). *The flow scales manual*. Morgantown: Publishers Graphics.

Jackson, S.A. et al. (2001). Relationships between flow, self-concept, psychological skills, and performance. *Journal of Applied Sports Psychology*, 13, 129–53.

James, N. (1978). *Woman alone*. London: Express Newspapers Ltd.

James, W. (1890). *The principles of psychology*. New York: H. Holt.

Janis, I.L. and Mann, L. (1977). *Decision making: A psychological analysis of conflict, choice, and commitment*. New York: Free Press.

Jaremka, L.M., Lindgren, M.E., and Kiecolt-Glaser, J.K. (2013). Synergistic relationships among stress, depression, and troubled relationships: Insights from psychoneuroimmunology. *Depression and Anxiety*, 30, 288–96.

Jeal, T. (2011). *Explorers of the Nile: The triumph and tragedy of a great Victorian adventure*. New Haven: Yale University Press.

Jeste, D.V. et al. (2010). Expert consensus on characteristics of wisdom: A Delphi method study. *The Gerontologist*, 50, 668–80.

Jha, A.P., Krompinger J., and Baime, M.J. (2007). Mindfulness training modifies subsystems of attention. *Cognitive, Affective and Behavioral Neuroscience*, 7, 109–19.

John, O.P., Robins, R.W., and Pervin, L.A. (Eds) (2008). *Handbook of personality: Theory and research* (3rd edition). New York: Guilford Press.

Johnson, M.B., Tenenbaum, G., and Edmonds, W.A. (2006). Adaptation to physically and emotionally demanding conditions: The role of deliberate practice. *High Ability Studies*, 17, 117–36.

Johnson, P.J. (2010). The roles of NASA, US astronauts and their families in long-duration missions. *Acta Astronautica*, 67, 561–71.

Johnson, P.L. et al. (2010). Sleep architecture changes during a trek from 1400 to 5000 m in the Nepal Himalaya. *Journal of Sleep Research*, 19, 148–56.

Jones, E., et al. (2004). Public panic and morale: A reassessment of civilian reactions during the Blitz and World War 2. *Journal of Social History*, 17, 463–79.

Jones, G., Hanton, S., and Connaughton, D. (2002). What is this thing called mental toughness? An investigation of elite sport performers. *Journal of Applied Sport Psychology*, 14, 205–18.

Joseph, S. (2011). *What doesn't kill us: The new psychology of posttraumatic growth*. London: Hachette Digital.

Jung, Y.-H. et al. (2010). The effects of mind-body training on stress reduction, positive affect, and plasma catecholamines. *Neuroscience Letters*, 479, 138–42.

Kabat-Zinn, J. (1982). An outpatient program in behavioural medicine for chronic pain patients based on the practice of mindfulness meditation: Theoretical considerations and preliminary results. *General Hospital Psychiatry*, 4, 33–47.

Kabat-Zinn, J. (1994). *Wherever you go, there you are: Mindfulness meditation in everyday life*. New York: Hyperion.

Kabat-Zinn, J. (2003). Mindfulness-based interventions in context: Past, present, and future. *Clinical Psychology: Science and Practice*, 10, 144–58.

Kabat-Zinn, J., Lipworth, L., and Burney, R. (1985). The clinical use of mindfulness meditation for the self-regulation of chronic pain. *Journal of Behavioral Medicine*, 8, 163–90.

Kahn, P.M. and Leon, G.R. (1994). Group climate and individual functioning in an all women's Antarctic expedition team. *Environment and Behavior*, 26, 669–97.

Kahneman, D. and Tversky, A. (1982). Variants of uncertainty. *Cognition*, 11, 143–57.

Kahneman, D. et al. (1993). When more pain is preferred to less: Adding a better end. *Psychological Science*, 4, 401–5.

Kamler, K. (2004). *Surviving the extremes: A doctor's journey to the limits of human endurance*. New York: St Martin's Press.

Kanas, N. (1998). Psychosocial issues affecting crews during long-duration international space missions. *Acta Astronautica*, 42, 339–61.

Kanas, N. and Manzey, D. (2008). *Space psychology and psychiatry* (2nd edition). Dordrecht: Springer.

Kanas, N., Weiss, D.S., and Marmar, C.R. (1996). Crew member interactions during a Mir space station simulation. *Aviation, Space, and Environmental Medicine, 67*, 969–75.

Kanas, N. et al. (2001). Psychosocial issues in space: Results from Shuttle/Mir. *Gravitational and Space Biology Bulletin, 14*, 35–45.

Kanas, N. et al. (2007). Crewmember and mission control personnel interactions during International Space Station missions. *Aviation, Space, and Environmental Medicine, 78*, 601–7.

Kaplan, S. (1995). The restorative benefits of nature: Toward an integrative framework. *Journal of Environmental Psychology, 15*, 169–82.

Kaplan, S. (2001). Meditation, restoration, and the management of mental fatigue. *Environment and Behavior, 33*, 480–506.

Karafantis, L. (2013). Sealab II and Skylab: Psychological fieldwork in extreme spaces. *Historical Studies in the Natural Sciences, 43*, 551–88.

Karkshan, E.M., Joharji, H.S., and Al-Harbi, N.N. (2002). Congenital insensitivity to pain in four related Saudi families. *Pediatric Dermatology, 19*, 333–5.

Kass, S.J., Vodanovich, S.J., and Callender, A. (2001). State-trait boredom: Relationship to absenteeism, tenure, and job satisfaction. *Journal of Business and Psychology, 16*, 317–27.

Kayes, D.C. (2006). *Destructive goal pursuit: The Mount Everest disaster.* New York: Palgrave Macmillan.

Keenan, B. (1993). *An evil cradling.* London: Vintage.

Kelly, A.D. and Kanas, N. (1992). Crewmember communication in space: A survey of astronauts and cosmonauts. *Aviation, Space, and Environmental Medicine, 63*, 721–6.

Kerr, J.H. (2007). Sudden withdrawal from skydiving: A case study informed by Reversal Theory's concept of protective frames. *Journal of Applied Sport Psychology, 19*, 337–51.

Killgore, W.D.S. (2007). Effects of sleep deprivation and morningness-eveningness traits on risk-taking. *Psychological Reports, 100*, 613–26.

King, J.H. (1878). Brief account of the sufferings of a detachment of United States Cavalry. *American Journal of Medical Science, 75*, 404–8.

King, L.A. et al. (1998). Resilience-recovery factors in post-traumatic stress disorder among female and male Vietnam veterans: Hardiness, postwar social support, and additional stressful life events. *Journal of Personality and Social Psychology, 74*, 420–34.

Kittinger, J. and Ryan, C. (2011). *Come up and get me: An autobiography of Colonel Joe Kittinger.* Albequerque: University of New Mexico Press.

Kjærgaard, A., Leon, G.R., and Fink, B.A. (in press). Personal challenges, communication processes, and team effectiveness in military special patrol teams operating in a polar environment. *Environment and Behavior.*

Kjærgaard, A. et al. (2013). Personality, personal values and growth in military special unit patrol teams operating in a polar environment. *Military Psychology, 25*, 13–22.

Kjellberg, A., Muhr P., and Skoldstrom, B. (1998). Fatigue after work in noise: An epidemiological survey study and three quasi-experimental field studies. *Noise Health, 1*, 47–55.

Klein, G. (1997). The current status of the naturalistic decision making framework. In R. Flin et al. (Eds) *Decision making under stress: Emerging themes and applications* (pp. 11–28). Aldershot: Ashgate.

Klein, G. (1998). *Sources of power: How people make decisions.* Cambridge, MA: The MIT Press.

Klein, G. and Crandall, B.W. (1995). The role of mental simulation in problem solving and decision making. In J. Caird et al. (Eds) *Local applications of the ecological approach to human-machine systems* (pp. 324–58). Hillsdale: L. Erlbaum Associates.

Knox-Johnston, R. (2004). *A world of my own.* London: Adlard Coles Nautical (originally published 1969).

Kolditz, T.A. (2007). *In extremis leadership: Leading as if your life depended on it*. San Francisco: Jossey-Bass.

Korchin, S. and Ruff, G.E. (1964). Personality characteristics of the Mercury astronauts. In G.H. Grosser et al. (Eds) *The threat of impending disaster* (pp. 203–4). Cambridge, MA: MIT Press.

Kouzes, J.M. and Posner, B.Z. (1992). The credibility factor: What people expect of leaders. In R.L. Taylor and W.E. Rosenbach (Eds) *Military leadership: In pursuit of excellence* (pp. 133–8). Boulder, CO: Westview.

Kozlowski, S.W. and Bell, B.S. (2003). Work groups and teams in organizations. In W.C. Borman, D.R. Ilgen, and R.J. Klimoski (Eds) *Handbook of psychology, volume 12: Industrial and organisational psychology* (2nd edition) (pp. 333–75). Hoboken: John Wiley and Sons.

Kozlowski, S.W. and Ilgen, D.R. (2006). Enhancing the effectiveness of work groups and teams. *Psychological Science in the Public Interest, 7*, 77–124.

Kozlowski, S.W. et al. (1999). Developing adaptive teams: A theory of compilation and performance across levels and time. In D.R. Ilgen and E.D. Pulakos (Eds) *The changing nature of performance: Implications for staffing, motivation, and development* (pp. 240–92). San Francisco: Jossey-Bass.

Kozlowski, S.W. et al. (2009). Developing adaptive teams: A theory of dynamic team leadership. In E. Salas, G.F. Goodwin, and C.S. Burke (Eds) *Team effectiveness in complex organizations: Cross-disciplinary perspectives and approaches* (pp. 113–55). Mahwah, NJ: Laurence Erlabaum Associates.

Krakauer, J. (1997). *Into thin air*. London: Macmillan.

Krantz, D.L. (1998). Taming chance: Social science and everyday narratives. *Psychological Inquiry, 9*, 87–94.

Kring, J.P. and Kaminski, M.A. (2011). Gender composition and crew cohesion during long-duration space missions. In D.A. Vakoch (Ed.) *Psychology of space exploration: Contemporary research in historical perspective* (pp. 125–41). Washington, DC: National Aeronautics and Space Administration.

Kuhnen, C.M. and Knutson, B. (2005). The neural basis of financial risk taking. *Neuron, 47*, 763–70.

Kull, R. (2008). *Solitude: Seeking wisdom in extremes*. Novato, CA: New World Library.

Kumar S., Soren S., and Chaudhury S. (2009). Hallucinations: Etiology and clinical implications. *Industrial Psychiatry Journal, 18*, 119–26.

Kwapil, T.R. (1998). Social anhedonia as a predictor of the development of schizophrenia-spectrum disorders. *Journal of Abnormal Psychology, 107*, 558–65.

Landreth, G. (2003). In Sverdrup's wake. *Ocean Navigator January/February*. Available at <http://www.oceannavigator.com/January-February-2003/In-Sverdrups-wake> accessed February 2014.

Langan-Fox, J. et al. (2001). Analyzing shared and team mental models. *International Journal of Industrial Ergonomics, 28*, 99–112.

Lasenby-Lessard, J. and Morrongiello, B.A. (2011). Understanding risk compensation in children: Experience with the activity and level of sensation seeking play a role. *Accident Analysis and Prevention, 43*, 1341–7.

Lau, H., Tucker, M.A., and Fishbein, W. (2010). Daytime napping: Effects on human direct associative and relational memory. *Neurobiology of Learning and Memory, 93*, 554–60.

Lazar, S.W. et al. (2005). Meditation experience is associated with increased cortical thickness. *NeuroReport, 16*, 1893–7.

Lazarus, R.S. (1966). *Psychological stress and the coping process*. New York: McGraw-Hill.

Lazarus, R.S. and Folkman, S. (1984). *Stress, appraisal, and coping*. New York: Springer.

Leary, M.R., Herbst, K.C., and McCrary, F. (2003a). Finding pleasure in solitary activities: Desire for aloneness or disinterest in social contact? *Personality and Individual Differences, 35,* 59–68.

Leary, M.R., Kowalski, R.M., and Smith, L. (2003b). Case studies of the school shootings. *Aggressive Behavior, 29,* 202–14.

Lee, C. and Longo, V.D. (2011). Fasting vs dietary restriction in cellular protection and cancer treatment: From model organisms to patients. *Oncogene, 30,* 3305–16.

Lee, C., Neighbors, C., and Woods, B.A. (2007). Marijuana motives: Young adults' reasons for using marijuana. *Addictive Behaviors, 32,* 1384–94.

Leino-Kilpi, H. et al. (2001). Privacy: A review of the literature. *International Journal of Nursing Studies, 38,* 663–71.

Leiter, M.P., Gascón, S., and Martínez-Jarreta, B. (2010). Making sense of work life: A structural model of burnout. *Journal of Applied Social Psychology, 40,* 57–75.

Leon, G. (2005). Men and women in space. *Aviation, Space, and Environmental Medicine, 76,* B84–8.

Leon, G.R., McNally, C., and Ben-Porath, Y.S. (1989). Personality characteristics, mood, and coping patterns in a successful North Pole expedition team. *Journal of Research in Personality, 23,* 162–79.

Leon, G.R. and Sandal, G.M. (2003). Women and couples in isolated extreme environments. *Acta Astronautica, 53,* 259–67.

Leon, G.R., et al. (2002). A 1-year, three-couple expedition as a crew analog for a Mars mission. *Environment and Behavior, 34,* 672–700.

Lester, J.T. (1983). Wrestling with the self on Mount Everest. *Journal of Humanistic Psychology, 23,* 31–41.

Lester, J.T. (2004). Spirit, identity, and self in mountaineering. *Journal of Humanistic Psychology, 44,* 86–100.

Lester, P.B. et al. (2010). Developing courage in followers: Theoretical and applied perspectives. In C.L. Pury and S.J. Lopez (Eds) *The psychology of courage: Modern research on an ancient virtue* (pp. 187–207). Washington, DC: American Psychological Association.

Levine, S. (1957). Infantile experience and resistance to physiological stress. *Science, 126,* 405.

Levine, S. (2005). Developmental determinants of sensitivity and resistance to stress. *Psychoneuroendocrinology, 30,* 939–46.

Lewitus, G.M. and Schwartz, M. (2009). Behavioral immunization: Immunity to self-antigens contributes to psychological stress resilience. *Molecular Psychiatry, 14,* 532–6.

Lieberman P. et al. (2005). Mount Everest: A space analogue for speech monitoring of cognitive deficits and stress. *Aviation and Space Environmental Medicine, 76,* B198–207.

Lindbergh, C.A. (1993). *The Spirit of St Louis*. St Paul, MN: Minnesota Historical Society Press (originally published 1953).

Linenger J.M. (2000). *Off the planet*. New York: McGraw-Hill.

Linley, P.A. (2003). Positive adaptation to trauma: Wisdom as both process and outcome. *Journal of Traumatic Stress, 16,* 601–10.

Lipman-Blumen, J. (2005). *The allure of toxic leaders: Why we follow destructive bosses and corrupt politicians—and how we can survive them*. Oxford: Oxford University Press.

Lipshitz, R. and Shaul, O.B. (1997). Schemata and mental models in recognition-primed decision making. In C.E. Zsambok and G. Klein (Eds) *Naturalistic decision making* (pp. 293–303). Mahwah, NJ: Lawrence Erlbaum Associates.

Lipshitz, R. et al. (2001). Taking stock of naturalistic decision making. *Journal of Behavioral Decision Making*, 14, 331–52.

Ljungberg, J.K. and Parmentier, F.B.R. (2010). Psychological effects of combined noise and whole-body vibration: A review and avenues for future research. *Proceedings of the Institution of Mechanical Engineers, Part D: Journal of Automobile Engineering*, 224, 1289–302.

Loewenstein, G. (1999). Because it is there: The challenge of mountaineering...for Utility Theory. *Kyklos*, 52, 315–43.

Logan, R.D. (1985). The 'flow experience' in solitary ordeals. *Journal of Humanistic Psychology*, 25, 79–89.

Long, C.R. et al. (2003). Solitude experiences: Varieties, settings, and individual differences. *Personality and Social Psychology Bulletin*, 29, 578–83.

Lopes Cardozo, B. et al. (2012). Psychological distress, depression, anxiety, and burnout among international humanitarian aid workers: A longitudinal study. *PLoS ONE*, 7, e44948.

Lorist, M.M. and Tops, M. (2003). Caffeine, fatigue, and cognition. *Brain and Cognition*, 53, 82–94.

Lowden, A. et al. (2010). Eating and shift work—effects on habits, metabolism, and performance. *Scandinavian Journal of Work, Environment and Health*, 36, 150–62.

Lutz, A. et al. (2009). Mental training enhances attentional stability: Neural and behavioral evidence. *The Journal of Neuroscience*, 29, 13418–27.

Luyster, F.S. et al. (2012). Sleep: A health imperative. *Sleep*, 35, 727–34.

Lyons, D.M. and Parker, K.J. (2007). Stress inoculation-induced indications of resilience in monkeys. *Journal of Traumatic Stress*, 20, 423–33.

Maass, P. (2002). Climbing lessons from the school of Tomaz Humar. *Outside Magazine*, June 2002.

Macfarlane, R. (2003). *Mountains of the mind*. London: Granta Books.

Macintyre, B. (2007). *Agent Zigzag*. London: Bloomsbury.

Mackersey, I. (1990). *Jean Batten: The Garbo of the skies*. London: Macdonald.

Maddi, S.R. and Khoshaba, D.M. (1994). Hardiness and mental health. *Journal of Personality Assessment*, 63, 265–74.

Maitland, S. (2008). *A book of silence*. London: Granta.

Manzey, D. and Lorenz, D. (1998). Mental performance during short-term and long-term space-flight. *Brain Research Reviews*, 28, 215–21.

Manzey, D., Lorenz, B., and Poljakov, V. (1998). Mental performance in extreme environments: Results from a performance monitoring study during a 438-day spaceflight. *Ergonomics*, 41, 537–59.

Markman, A.B. and Gentner, D. (2001). Thinking. *Annual Review of Psychology*, 52, 223–47.

Martha, C. and Laurendeau, J. (2010). Are perceived comparative risks realistic among high-risk sports participants? *International Journal of Sport and Exercise Psychology*, 8, 129–46.

Martin, P. (1997). *The sickening mind: Brain, behaviour, immunity and disease*. London: HarperCollins.

Martin, P. (2002). *Counting sheep: The science and pleasures of sleep and dreams*. London: Harper Collins.

Martin, P. (2005). *Making happy people: The nature of happiness and its origins in childhood*. London: Fourth Estate.

Martin, P. (2008). *Sex, drugs and chocolate: The science of pleasure*. London: Fourth Estate.

Martin, P. and Bateson, P. (2007). *Measuring behaviour: An introductory guide* (3rd edition). Cambridge: Cambridge University Press.

Martin, R.A. (2001). Humor, laughter and physical health: Methodological issues and research findings. *Psychological Bulletin*, 127, 504–19.

Masicampo, E.J. and Baumeister, R.F. (2008). Toward a physiology of dual-process reasoning and judgment: Lemonade, willpower, and expensive rule-based analysis. *Psychological Science*, 19, 255–60.

Maslach, C. and Leiter, M.P. (1997). *The truth about burnout.* San Francisco: Jossey-Bass.

Mason, M.F. et al. (2007). Wandering minds: The default network and stimulus-independent thought. *Science*, 315, 393–5.

Mathwick, C. and Rigdon, E. (2004). Play, flow, and the online search experience. *Journal of Consumer Research*, 31, 324–32.

Mattoni, R.H. and Sullivan, G.H. (1962). *Sanitation and personal hygiene during aerospace missions.* Available at <http://www.dtic.mil/dtic/tr/fulltext/u2/283841.pdf> accessed February 2014.

Mattson, M.P. (2008). Dietary factors, hormesis and health. *Ageing Research Reviews*, 7, 43–8.

Mattson, M.P. and Wan, R. (2005). Beneficial effects of intermittent fasting and caloric restriction on the cardiovascular and cerebrovascular systems. *Journal of Nutritional Biochemistry*, 16, 129–37.

Mauri, M. et al. (2011). Why is Facebook so successful? Psychophysiological measures describe a core flow state while using Facebook. *Cyberpsychology, Behavior, and Social Networking*, 14, 723–31.

McCay, C.M., Crowell, M.F., and Maynard, L.A. (1935). The effect of retarded growth upon the length of lifespan and upon ultimate body size. *Journal of Nutrition*, 10, 63–79.

McCormick, I.A., et al. (1985). A psychometric study of stress and coping during the International Biomedical Expedition to the Antarctic (IBEA). *Journal of Human Stress*, 11, 150–6.

McCorristine, S. (2010). The supernatural Arctic: An exploration. *Nordic Journal of English Studies*, 9, 47–70.

McGurk, D. and Castro, C.A. (2010). Courage in combat. In C.L.S. Pury and S.J. Lopez (Eds) *The psychology of courage* (pp. 167–85). Washington, DC: American Psychological Association.

McKenna, B.S. et al. (2007). The effects of one night of sleep deprivation on known-risk and ambiguous-risk decisions. *Journal of Sleep Research*, 16, 245–52.

McKnight, D.H., Cummings, L.L., and Chervany, N.L. (1998). Initial trust formation in new organizational relationships. *Academy of Management Review*, 23, 473–90.

McMurray, K.F. (2001). *Deep descent: Adventure and death diving the Andrea Doria.* New York: Touchstone.

Mednick, S.C. et al. (2002). The restorative effect of naps on perceptual deterioration. *Nature Neuroscience*, 5, 677–81.

Mednick, S.C. et al. (2008). Comparing the benefits of caffeine, naps and placebo on verbal, motor and perceptual memory. *Behavioural Brain Research*, 193, 79–86.

Meeks, T.W. and Jeste, D.V. (2009). Neurobiology of wisdom: An overview. *Archives of General Psychiatry*, 66, 355–65.

Megdal, S.P. et al. (2005). Night work and breast cancer risk: A systematic review and meta-analysis. *European Journal of Cancer*, 41, 2023–32.

Meichenbaum, D.H. and Deffenbacher, J.L. (1988). Stress inoculation training. *The Counselling Psychologist*, 16, 69–90.

Mercer, K.B. and Eastwood, J.D. (2010). Is boredom associated with problem gambling behaviour? It depends on what you mean by 'boredom'. *International Gambling Studies*, 10, 91–104.

Meyer, J.P. and Allen, N.J. (1991). A three-component conceptualization of organizational commitment. *Human Resource Management Review*, 1, 61–89.

Meyer, J.P. et al. (2002). Affective, continuance, and normative commitment to the organization: A meta-analysis of antecedents, correlates, and consequences. *Journal of Vocational Behavior*, 61, 20–52.

Mignot, E. (2013). The perfect hypnotic? *Science*, 340, 36–8.

Miller, J.J., Fletcher, K., and Kabat-Zinn, J. (1995). Three-year follow-up and clinical implications of a mindfulness meditation-based stress reduction intervention in the treatment of anxiety disorders. *General Hospital Psychiatry*, 17, 192–200.

Mills, J.N. (1964). Circadian rhythms during and after three months in solitude underground. *Journal of Physiology*, 174, 217–31.

Milner, C.E. and Cote, K.A. (2009). Benefits of napping in healthy adults: Impact of nap length, time of day, age, and experience with napping. *Journal of Sleep Research*, 18, 272–81.

Milnes Walker, N. (1972). *When I put out to sea*. London: Pan Books Ltd.

Minde, J.K. (2006). Norrbottnian congenital insensitivity to pain. *Acta Orthopaedica Supplementum*, 77, 2–32.

Mischel, W. et al. (2011). 'Willpower' over the life span: Decomposing self-regulation. *Social Cognitive and Affective Neuroscience*, 6, 252–6.

Mishra, V. (2006). Stress, anxiety and loneliness among 20th Indian expeditioners at Antarctica during summer. *Indian Ministry of Earth Sciences, Technical Publication No. 18*, 233–41.

Mitchell, D.E. (2008). *The way of the explorer: An Apollo astronaut's journey through the material and mystical worlds*. Franklin Lakes, NJ: The Career Press.

Mitchell, S.J. et al. (2007). Fatal respiratory failure during a 'technical' rebreather dive at extreme pressure. *Aviation, Space, and Environmental Medicine*, 78, 81–6.

Mocellin, J. et al. (1991). Levels of anxiety in polar environments. *Journal of Environmental Psychology*, 11, 265–75.

Mocellin, J. (1995). Levels of anxiety aboard two expeditionary ships. *The Journal of General Psychology*, 122, 317–24.

Mocellin, J. and Suedfeld, P. (1991). Voices from the ice: Diaries of Polar explorers. *Environment and Behavior*, 23, 704–22.

Moitessier, B. (1995). *The long way*. Dobbs Ferry, NY: Sheridan House Inc. (originally published 1971).

Monasterio, E. (2006). Adventure sports in New Zealand: Dangerous and costly recklessness or valuable health-promoting activity? Be careful to judge. *The New Zealand Medical Journal*, 119, 5–7.

Morgan, C.A. et al. (2000). Plasma neuropeptide-Y concentrations in humans exposed to military survival training. *Biological Psychiatry*, 47, 902–9.

Morgan, W.P. (1995). Anxiety and panic in recreational SCUBA divers. *Sports Medicine*, 20, 398–421.

Morrell, M. and Capparell, S. (2003). *Shackleton's way*. London: Nicholas Brealey Publishing.

Morris, J. (1958). *Coronation Everest*. London: Faber and Faber.

Morton P.A. (2003). The hypnotic belay in Alpine mountaineering: The use of self-hypnosis for the resolution of sports injuries and for performance enhancement. *American Journal of Clinical Hypnosis*, 46, 45–51.

Moskovitz, C. (2011). Remembering 9/11: An astronaut's painful view from space. Available at <http://www.space.com/12877-september-11-space-station-astronaut-culbertson.html> accessed February 2014.

Mountfield, D. (1974). *A history of polar exploration*. New York: Dial Press.

Nansen, F. (2008). *Farthest North*. New York: Skyhorse Publishing (originally published 1897).

Neave, N. et al. (2001). Water ingestion improves subjective alertness, but has no effect on cognitive performance in dehydrated healthy young volunteers. *Appetite, 37*, 255–6.

Nelson, T.O. et al. (1990). Cognition and metacognition at extreme altitudes on Mount Everest. *Journal of Experimental Psychology: General, 119*, 367–74.

Nettle, D. (2007). *Personality: What makes you the way you are*. Oxford: Oxford University Press.

Newby, E. (1956). *The last grain race*. London: Secker and Warburg.

Newby-Clark, I.R. et al. (2000). People focus on optimistic scenarios and disregard pessimistic scenarios while predicting task completion times. *Journal of Experimental Psychology: Applied, 6*, 171–82.

Nezlek, J.B. et al. (2012). Ostracism in everyday life. *Group Dynamics: Theory, Research, and Practice, 16*, 91–104.

Nichols, P. (1997). *A voyage for madmen*. London: Profile Books.

Norton, T. (2000). *Stars beneath the sea: The extraordinary lives of the pioneers of diving*. London: Arrow.

Noyce, W. (1958). *The springs of adventure*. London: John Murray.

Nunn, W.C. (1940). Eighty-six hours without water on the Texas Plains. *The Southwestern Historical Quarterly, 43*, 356–64.

Nussbaumer-Ochsner, Y. et al. (2012). Effect of short-term acclimatization to high altitude on sleep and nocturnal breathing. *Sleep, 35*, 419–23.

O'Brien, M. (1999). Shuttle astronaut taken off crew for ISS mission. CNN.com, 8 September. Available at <http://edition.cnn.com/TECH/space/9909/08/astronaut.removed/index.html> accessed February 2014.

Oaten, M., Stevenson, R.J., and Case, T.I. (2009). Disgust as a disease-avoidance mechanism. *Psychological Bulletin, 135*, 303–21.

Oaten, M. et al. (2008). The effects of ostracism on self-regulation in the socially anxious. *Journal of Social and Clinical Psychology, 27*, 471–504.

Oberg, J.E. and Oberg, A.R. (1986). *Pioneering space*. New York: McGraw Hill.

Ohayon, M.M. (2000). Prevalence of hallucinations and their pathological associations in the general population. *Psychiatry Research, 97*, 153–64.

Olatunji, B.O., et al. (2007). The disgust scale: Item analysis, factor structure, and suggestions for refinement. *Psychological Assessment, 19*, 281.

Orasanu, J. (1997). Stress and naturalistic decision making: Strengthening the weak links. In R. Flin et al. (Eds) *Decision making under stress*. Aldershot: Ashgate.

Orasanu, J. (2005). Crew collaboration in space: A naturalistic decision-making perspective. *Aviation, Space, and Environmental Medicine, 76*, B154–63.

Orasanu, J. (2010). Flight crew decision-making. In B. Kanki, R. Helmreich, and J. Anca (Eds) *Crew resource management* (pp. 147–79). San Diego, CA: Academic Press.

Oswald, I. (1974). *Sleep* (3rd edition). Harmondsworth: Penguin

Owen, M. (2012). *No easy day: The autobiography of a Navy SEAL*. London: Penguin.

Padilla, A., Hogan, R., and Kaiser, R.B. (2007). The toxic triangle: Destructive leaders, susceptible followers, and conducive environments. *The Leadership Quarterly, 18*, 176–94.

Palinkas, L.A. (1991). Effects of physical and social environments on the health and well-being of Antarctic winter-over personnel. *Environment and Behavior, 23*, 782–99.

Palinkas, L.A. et al. (2004a). Cross-cultural differences in psychosocial adaptation to isolated and confined environments. *Aviation, Space and Environmental Medicine, 75*, 973–80.

Palinkas, L.A. et al. (2004b). Incidence of psychiatric disorders after extended residence in Antarctica. *International Journal of Circumpolar Health*, 63, 157–68.

Palinkas, L.A. and Suedfeld, P. (2008). Psychological effects of polar expeditions. *The Lancet*, 371(9607), 153–63.

Palinkas, L.A. et al. (2000). Predictors of behavior and performance in extreme environments: The Antarctic space analogue program. *Aviation, Space, and Environmental Medicine*, 71, 619–25.

Pallesen, S. et al. (2007). Prevalence and risk factors of subjective sleepiness in the general adult population. *Sleep*, 30, 619–24.

Palmai, G. (1963). Psychological observations on an isolated group in Antarctica. *The British Journal of Psychiatry*, 109, 364–70.

Pekrun, R. et al. (2010). Boredom in achievement settings: Exploring control-value antecedents and performance outcomes of a neglected emotion. *Journal of Educational Psychology*, 102, 531–49.

Peldszus, R. et al. (2014). The perfect boring situation—addressing the experience of monotony during crewed deep space missions through habitability design. *Acta Astronautica*, 94, 262–76.

Pemberton, J. (2006). Medical experiments carried out in Sheffield on conscientious objectors to military service during the 1939–45 war. *International Journal of Epidemiology*, 35, 556–8.

Pennebaker, J.W. (1997). Writing about emotional experiences as a therapeutic process. *Psychological Science*, 8, 162–6.

Peplau, L.A. and Perlman, D. (1982). Perspectives on loneliness. In L.A. Peplau, and D. Perlman (Eds) *Loneliness: A sourcebook of current theory, research and therapy* (pp. 1–18). New York: John Wiley.

Perchonok, M. and Douglas, G. (2009). Risk factor of inadequate food system. In J.C. McPhee and J.B. Charles (Eds) *NASA SP-2009–3405: Human health and performance risks of space exploration missions. Evidence reviewed by the NASA Human Research Program* (pp. 295–316). Houston, TX: NASA.

Perlman, D. and Peplau, L.A. (1984). Loneliness research: A survey of empirical findings. In L.A. Peplau and S. Goldston (Eds) *Preventing the harmful consequences of severe and persistent loneliness* (pp. 13–46). Rockville, MD: National Institute of Mental Health.

Pervin, L.A. and John, O.P. (1997). *Personality: Theory and research* (7th Edn). New York: John Wiley and Sons.

Petri, N.M. (2003). Change in strategy of solving psychological tests: Evidence of nitrogen narcosis in shallow air-diving. *Undersea and Hyperbaric Medicine*, 30, 293–303.

Piantadosi, C.A. (2003). *The biology of human survival: Life and death in extreme environments.* New York,: Oxford University Press.

Plous, S. (1993). *The psychology of judgment and decision making.* New York: McGraw-Hill.

Popkin, B.M., D'Anci, K.E., and Rosenberg, I.H. (2010). Water, hydration, and health. *Nutrition Reviews*, 68, 439–58.

Posner, B.Z. and Kouzes, J.M. (1988). Relating leadership and credibility. *Psychological Reports*, 63, 527–30.

Powter, G. (2006). *Strange and dangerous dreams.* Seattle, WA: The Mountaineers Books.

Poynter, J. (2006). *The human experiment: Two years and twenty minutes inside Biosphere 2.* New York: Basic Books.

Pritchard, P. (1999). *The Totem Pole: And a whole new adventure.* London: Constable and Co.

Pritchard, P. (2005). *The longest climb: Back from the abyss.* London: Robinson.

Pury, C.L., Lopez, S.J., and Key-Roberts, M. (2010). The future of courage research. In C.L. Pury and S.J. Lopez (Eds) *The psychology of courage: Modern research on an ancient virtue* (pp. 229–35). Washington, DC: American Psychological Association.

Rachman (1990). *Fear and courage* (2nd Edn). New York: WH Freeman and Co.

Rachman, S. (1994). The over-prediction of fear: A review. *Behaviour Research and Therapy, 32*, 683–90.

Radcliffe, N.M. and Klein, W.M.P. (2002). Dispositional, unrealistic, and comparative optimism: Differential relations with the knowledge and processing of risk information and beliefs about personal risk. *Personality and Social Psychological Bulletin, 28*, 836–46.

Radloff, R. and Helmreich, R.L. (1968). *Groups under stress: Psychological research in SEALAB II.* Englewood Cliffs, NJ: Prentice-Hall.

Rajaratnam, S.M.W. and Arendt, J. (2001). Health in a 24-h society. *The Lancet, 358*, 999–1005.

Ramel, W. et al. (2004). The effects of mindfulness meditation on cognitive processes and affect in patients with past depression. *Cognitive Therapy and Research, 28*, 433–55.

Rapp, G.C. (2008). The wreckage of recklessness. *Washington University Law Review, 86*, 111–80.

Rassovsky, Y. and Kushner, M.G. (2003). Carbon dioxide in the study of panic disorder: Issues of definition, methodology, and outcome. *Anxiety Disorders, 17*, 1–32.

Reason, J. (2008). *The human contribution.* Farnham: Ashgate Publishing Ltd.

Reed, G.E. (2004). Toxic leadership. *Military Review, 84*, 67–71.

Reider, R. (2010). *Dreaming the biosphere.* Albuquerque: University Of New Mexico Press.

Remick, A.K., Polivy, J., and Pliner, P. (2009). Internal and external moderators of the effect of variety on food intake. *Psychological Bulletin, 135*, 434–51.

Rihel, J. and Schier, A.F. (2013). Sites of action of sleep and wake drugs: Insights from model organisms. *Current Opinion in Neurobiology, 23*, 831–40.

Rip, B., Fortin, S., and Vallerand, R.J. (2006). The relationship between passion and injury in dance students. *Journal of Dance Medicine and Science, 10*, 14–20.

Ritsher, J.B. (2005). Cultural factors and the International Space Station. *Aviation, Space, and Environmental Medicine, 76*, 135–44.

Ritsher, J.B., Kanas, N., and Saylor, S. (2005). Maintaining privacy during psychosocial research on the International Space Station. *Journal of Human Performance in Extreme Environments, 8(1)*, Article 3. Available at <http://docs.lib.purdue.edu/jhpee/vol8/iss1/3> accessed February 2014.

Ritsher, J.B., et al. (2007). Psychological adaptation and salutogenesis in space: Lessons from a series of studies. *Acta Astronautica, 60*, 336–40.

Rivolier, J. et al. (1988). *Man in the Antarctic: The scientific work of the International Biomedical Expedition to the Antarctic (IBEA).* New York: Taylor and Francis.

Rivolier, J., Cazes, G., and McCormick, I. (1991). The International Biomedical Expedition to the Antarctic: Psychological evaluations of the field party. In A.A. Harrison, Y.A. Clearwater, and C.P. McKay (Eds) *From Antarctica to outer space* (pp. 283–90). New York: Springer-Verlag.

Roach, M. (2010). *Packing for Mars: The curious science of life in space.* Oxford: Oneworld.

Roberts, D. (2013). *Alone on the ice: The greatest survival story in the history of exploration.* New York: W.W. Norton.

Robinson, M.E., Staud, R., and Price, D.D. (2013). Pain measurement and brain activity: Will neuroimages replace pain ratings? *The Journal of Pain, 14*, 323–7.

Robson, C. (2011). *Real world research* (3rd Edn). Chichester: Wiley.

REFERENCES

Rogers, P.J. et al. (2001). A drink of water can improve or impair mental performance depending on small differences in thirst. *Appetite*, 36, 57–8.

Rolls, B.J. et al. (1980). Thirst following water deprivation in humans. *American Journal of Physiology—Regulatory Integrative and Comparative Physiology*, 239, R476–82.

Rose, D. and Douglas, E. (2000). *Regions of the heart*. London, Penguin.

Rosnet, E. et al. (2004). Mixed-gender groups: Coping strategies and factors of psychological adaptation in a polar environment. *Aviation, Space, and Environmental Medicine*, 75 (7, Supplement), C10–13.

Rowland, G.L., Franken, R.E., and Harrison, K. (1986). Sensation seeking and participation in sporting activities. *Journal of Sport Psychology*, 8, 212–22.

Rubin, K.H., Coplan, R.J., and Bowker, J.C. (2009). Social withdrawal in childhood. *Annual Review of Psychology*, 60, 141–71.

Ruff, G. and Korchin, S. (1964). Psychological responses of the Mercury astronauts to stress. In G.H. Grosser et al. (Eds) *The threat of impending disaster* (pp. 208–20). Cambridge, MA: MIT Press.

Ruff, G.E. (2010). Commentary. *Aviation, Space, and Environmental Medicine*, 81, 157.

Rufus, A.S. (2003). *Party of one: The loner's manifesto*. Cambridge, MA: Da Capo Press.

Russell, B. (1950). What desires are politically important? *Nobel Lecture*, 11 December. Available at <http://www.nobelprize.org/nobel_prizes/literature/laureates/1950/russell-lecture.html> accessed February 2014.

Russell, D.W. (1996). UCLA Loneliness Scale (version 3): Reliability, validity, and factor structure. *Journal of Personality Assessment*, 66, 20–40.

Russell, D.W. et al. (2012): Is loneliness the same as being alone? *The Journal of Psychology: Interdisciplinary and Applied*, 146, 7–22.

Ryan, C. (1995). *The pre-astronauts*. Annapolis, MD: Naval Institute Press.

Ryan, C. (2003). *Magnificent failure: Freefall from the edge of space*. Washington, DC: Smithsonian Institution Press.

Sacks, O. (2012). *Hallucinations*. London: Picador.

Salas, E. et al. (2012). The science of training and development in organizations: What matters in practice. *Psychological Science in the Public Interest*, 13, 74–101.

Sandal, G.M. (2004). Culture and tension during an International Space Station simulation: Results from SFINCSS'99. *Aviation, Space, and Environmental Medicine*, 75, C44–51.

Sandal, G.M., Hege, H.B., and van de Vijver, F.J.R. (2011). Personal values and crew compatibility: Results from a 105 days simulated space mission. *Acta Astronautica*, 69, 141–9.

Sandoval, L. et al. (2012). *Perspectives on asthenia in astronauts and cosmonauts: Review of the international research literature*. Houston, TX: NASA.

Sanna, L.J. et al. (2005). The hourglass is half full or half empty: Temporal framing and the group planning fallacy. *Group Dynamics: Theory, Research, and Practice*, 9, 173–88.

Santy, P.A. (1994). *Choosing the right stuff: The psychological selection of astronauts and cosmonauts*. Westport, CT: Praeger.

Santy, P.A. et al. (1988). Analysis of sleep on Shuttle missions. *Aviation, Space, and Environmental Medicine*, 59, 1094–7.

Sarramon, C. et al. (1999). Addiction and personality traits: Sensation seeking, anhedonia, impulsivity. *Encephale*, 25, 569–75.

Sarris, A. (2007). Antarctic culture: 50 years of Antarctic expeditions. *Aviation, Space, and Environmental Medicine*, 78, 886–92.

Schernhammer, E.S. et al. (2001). Night-shift work and risk of colorectal cancer in the nurses' health study. *Journal of the National Cancer Institute, 93,* 825–8.

Schneider, B. and Reichers, A.E. (1983). On the etiology of climates. *Personnel Psychology, 36,* 19–39.

Schoenfeld, A.H. (1992). Learning to think mathematically: Problem solving, metacognition, and sense making in mathematics. In D.A. Grouws (Ed.) *Handbook of research on mathematics teaching and learning* (pp. 334–70). New York: Macmillan.

Schoonover, J. (Ed.) (2007). *Adventurous dreams, adventurous lives.* Calgary: Rocky Mountain Books Incorporated.

Scott, G.D. and Gendreau, P. (1969). Psychiatric implications of sensory deprivation in a maximum security prison. *Canadian Psychiatric Association Journal, 14,* 337–41.

Scott, P. (2005). No bath time. *Scientific American, 292,* 26–7.

Seery, M.D. (2011). Resilience: A silver lining to experiencing adverse life events? *Current Directions in Psychological Science, 20,* 390–4.

Seery, M.D., et al. (2013). An upside to adversity? Moderate cumulative lifetime adversity is associated with resilient responses in the face of controlled stressors. *Psychological Science, 24,* 1181–9.

Segrin, C. and Kinney, T. (1995). Social skills deficits among the socially anxious: Rejection from others and loneliness. *Motivation and Emotion, 19,* 1–24.

Seib, H.M. and Vodanovich, S.J. (1998). Cognitive correlates of boredom proneness: The role of private self-consciousness and absorption. *The Journal of Psychology, 132,* 642–52.

Serfaty, D., Entin, E.E., and Johnston, J.H. (1998). Team coordination training. In J.A. Cannon-Bowers and E. Salas (Eds) *Making decisions under stress: Implications for individual and team training* (pp. 221–45). Washington, DC: American Psychological Association.

Sexton, J.B. and Helmreich, R.L. (2000). Analyzing cockpit communications: The links between language, performance, error, and workload. *Journal of Human Performance in Extreme Environments, 5(1),* Article 6. Available at <http://docs.lib.purdue.edu/jhpee/vol5/iss1/6/> accessed February 2014.

Shackleton, E.H. (1920). *South: The story of Shackleton's last expedition, 1914–1917.* New York: Macmillan.

Shalev, S. (2008). *A sourcebook on solitary confinement.* London: London School of Economics.

Shaw, J.B., Erickson, A., and Harvey, M. (2011). A method for measuring destructive leadership and identifying types of destructive leaders in organisations. *The Leadership Quarterly, 22,* 575–90.

Sheard, M. (2010). *Mental toughness: The mindset behind sporting achievement.* London: Routledge.

Shepperd, J.A. (1993). Productivity loss in performance groups: A motivational analysis. *Psychological Bulletin, 113,* 67–81.

Shepperd, J.A. et al. (2013). Taking stock of unrealistic optimism. *Perspectives on Psychological Science, 8,* 395–411.

Shuffler, M.L., Diaz Granados, D., and Salas, E. (2011). There's a science for that: Team development interventions in organizations. *Current Directions in Psychological Science, 20,* 365–72.

Sicard, B., Jouve, E., and Blin, O. (2001). Risk propensity assessment in military special operations. *Military Medicine, 166,* 871–4.

Siegel, R.K. (1984). Hostage hallucinations: Visual imagery induced by isolation and life-threatening stress. *The Journal of Nervous and Mental Disease, 172,* 264–72.

Siffre, M. (1964). *Beyond time.* New York: McGraw-Hill.

Siffre, M. (1975). Six months alone in a cave. *National Geographic*, 147, 426–35.

Signal, T.L. et al. (2012). Duration of sleep inertia after napping during simulated night work and in extended operations. *Chronobiology International*, 29, 769–79.

Silvia, P.J. and Kwapil, T.R. (2011). Aberrant asociality: How individual differences in social anhedonia illuminate the need to belong. *Journal of Personality*, 79, 1315–32.

Simonet, S. and Wilde, G.J. (1997). Risk: Perception, acceptance and homeostasis. *Applied Psychology*, 46, 235–52.

Simons, D.G. and Schanche, D.A. (1960). *Man high*. New York: Doubleday.

Simonton, D.K. (2003). Qualitative and quantitative analyses of historical data. *Annual Review of Psychology*, 54, 617–40.

Simpson, J. (1997). *Touching the void*. London: Vintage.

Simpson, J. (2003). *The beckoning silence*. London: Vintage.

Slack, K. et al. (2009). Risk of behavioral and psychiatric conditions. In J.C. Mcphee and J.B. Charles (Eds) *Human health and performance risks of space exploration missions* (pp. 3–45). Houston, TX: NASA.

Slocum, J. (2006). *Sailing alone around the world*. Bedford, MA: Applewood Books (originally published 1900).

Smallwood, J., McSpadden, M., and Schooler, J.W. (2008). When attention matters: The curious incident of the wandering mind. *Memory and Cognition*, 36, 1144–50.

Smallwood, J. and Schooler, J.W. (2006). The restless mind. *Psychological Bulletin*, 132, 946–58.

Smallwood, J. et al. (2009). Shifting moods, wandering minds: Negative moods lead the mind to wander. *Emotion*, 9, 271–6.

Smith, A. (2005). *Moondust: In search of the men who fell to Earth*. London: Bloomsbury.

Smith, A. (2012). An update on noise and performance: Comment on Szalma and Hancock (2011). *Psychological Bulletin*, 138, 1262–8.

Smith, A. and Williams, K.D. (2004). RU there? Ostracism by cell phone text messages. *Group Dynamics: Theory, Research, and Practice*, 8, 291.

Smith, A.P. (2012). Twenty-five years of research on the behavioural malaise associated with influenza and the common cold. *Psychoneuroendocrinology*, 38, 744–51.

Smith, M. (2002). *I am just going outside: Captain Oates—Antarctic tragedy*. Staplehurst: Spellmount.

Sommers, J. and Vodanovich, S.J. (2000). Boredom proneness: Its relationship to psychological- and physical-health symptoms. *Journal of Clinical Psychology*, 56, 149–55.

Sontag, S. and Drew, C. (1999). *Blind man's bluff: The untold story of Cold War submarine espionage*. London: Arrow.

Southwick, S.M. and Charney, D.S. (2012). *Resilience: The science of mastering life's greatest challenges*. Cambridge: Cambridge University Press.

Spacks, P.M. (1995). *Boredom: The literary history of a state of mind*. Chicago: University of Chicago Press.

Speca, M. et al. (2000). A randomized, wait-list controlled clinical trial: The effect of a mindfulness-based stress reduction program on mood and symptoms of stress in cancer outpatients. *Psychosomatic Medicine*, 62, 613–22.

Speke, J.H. (1864). *What led to the source of the Nile*. London: William Blackwood and Sons.

Spielberger, C. (1972). *Anxiety: Current trends in research*. London: Academic Press.

Spufford, F. (2003). *I may be some time: Ice and the English imagination*. London: Faber and Faber.

Stajkovic, A.D. and Luthans, F. (1998). Self-efficacy and work-related performance: A meta-analysis. *Psychological Bulletin*, 124, 240–61.

Stanton, N.A. and Pinto, M. (2000). Behavioural compensation by drivers of a simulator when using a vision enhancement system. *Ergonomics, 43*, 1359–70.

Steele, J.P. (2011). *Antecedents and consequences of toxic leadership in the US Army: A two year review and recommended solutions (technical report 2011–3)*. Kansas: Center for Army Leadership.

Stein, G.M. (2011). An Arctic execution: Private Charles B. Henry of the United States Lady Franklin Bay Expedition 1881–84. *Arctic, 64*, 399–412.

Sternberg, R.J. (2001). Why schools should teach for wisdom: The balance theory of wisdom in educational settings. *Educational Psychologist, 36*, 227–45.

Stevens, R.G. (2009). Light-at-night, circadian disruption and breast cancer: Assessment of existing evidence. *International Journal of Epidemiology, 38*, 963–70.

Stogdill, R.M. (1974). *Handbook of leadership: A survey of theory and research*. New York: Free Press.

Stone, W.C. (2004). Deep/underwater cave environments. In S.J. Dick and K.L. Cowing (Eds) *Risk and exploration: Earth, sea, and the stars. NASA Administrator's symposium September 26–29, 2004, Naval Postgraduate School Monterey, California* (pp. 70–6). Washington, DC: NASA.

Storr, A. (1988). *Solitude*. London: HarperCollins.

Stote, K.S. et al. (2007). A controlled trial of reduced meal frequency without caloric restriction in healthy, normal-weight, middle-aged adults. *American Journal of Clinical Nutrition, 85*, 981–8.

Stuster J. (1996). *Bold endeavors: Lessons from space and polar exploration*. Annapolis, MD: Naval Institute Press.

Stuster, J. (2010). *Behavioral issues associated with long duration space expeditions: Review and analysis of astronaut journals*. NASA Technical Manuscript, 216130. Houston, TX: NASA.

Stuster, J., Bachelard, C., and Suedfeld, P. (1999). *In the wake of the astrolabe: Review and analysis of diaries maintained by the leaders and physicians at French remote-duty stations*. Technical Report 1159 for the National Aeronautics and Space Administration. Santa Barbara, CA: Anacapa Sciences, Inc.

Suedfeld, P. (2010). Mars: Anticipating the next great exploration. Psychology, culture and camaraderie. *Journal of Cosmology, 12*, 1–8.

Suedfeld, P. and Bow, R.A. (1999). Health and therapeutic applications of chamber and flotation restricted environmental stimulation therapy (REST). *Psychology and Health, 14*, 545–66.

Suedfeld, P. and Coren, S. (1989). Perceptual isolation, sensory deprivation, and rest: Moving introductory psychology texts out of the 1950s. *Canadian Psychology/Psychologie Canadienne, 30*, 17–29.

Suedfeld, P. and Mocellin, J.S.P. (1987). The sensed presence in unusual environments. *Environment and Psychology, 19*, 33–52.

Suedfeld, P. and Steel, G.D. (2000). The environmental psychology of capsule habitats. *Annual Review of Psychology, 51*, 227–53.

Suedfeld, P. and Weiszbeck, T. (2004). The impact of outer space on inner space. *Aviation, Space, and Environmental Medicine, 75*, C6–9.

Suedfeld, P., Wilk, K.E., and Cassel, L. (2011). Flying with strangers: Postmission reflections of multinational space crews. In D.A. Vakoch (Ed.) *Psychology of space exploration in historical perspective* (pp. 143–75). Washington, DC: NASA.

Svendsen, L. (2005). *A philosophy of boredom*. (Trans. J. Irons.) London: Reaktion Books.

Svenson, O. (1981). Are we all less risky and more skilful than our fellow drivers? *Acta Psychologica, 47*, 143–8.

Szalma, J.L. and Hancock, P.A. (2011). Noise effects on human performance: A meta-analytic synthesis. *Psychological Bulletin, 137*, 682–707.

Tabor, J.M. (2011). *Blind descent: The quest to discover the deepest place on earth.* London: Random House.

Tajfel, H. (Ed.) (1982). *Social identity and intergroup relations.* Cambridge: Cambridge University Press.

Takahashi, M. and Arito, H. (2000). Maintenance of alertness and performance by a brief nap after lunch under prior sleep deficit. *Sleep, 23*, 813–19.

Taleb, N.N. (2012). *Antifragile.* London: Allen Lane.

Tang, Y-Y. et al. (2007). Short-term meditation training improves attention and self-regulation. *Proceedings of the National Academy of Sciences USA, 104*, 17152–6.

Taylor, A.J.W. (1991). The research program of the International Biomedical Expedition to the Antarctic (IBEA) and its implications for research in outer space. In A.A. Harrison, Y.A. Clearwater, and C.P. McKay (Eds) *From Antarctica to outer space* (pp. 43–56). New York: Springer-Verlag.

Taylor, S.E. (2011). Affiliation and stress. In S. Folkman (Ed.) *Oxford handbook of stress, health, and coping* (pp. 86–100). New York: Oxford University Press.

Tedeschi, R.G. and Calhoun, L.G. (2004). Post-traumatic growth: Conceptual foundations and empirical evidence. *Psychological Inquiry, 15*, 1–18.

Teese, R. and Bradley, G. (2008). Predicting recklessness in emerging adults: A test of a psychosocial model. *The Journal of Social Psychology, 148*, 105–26.

Teunisse, R.J. et al. (1995). The Charles Bonnet syndrome: A large prospective study in The Netherlands. *The British Journal of Psychiatry, 166*, 254–7.

Teunisse R.J. et al. (1996). Visual hallucinations in psychologically normal people: Charles Bonnet's syndrome. *The Lancet, 347*, 794–7.

Thesiger, W. (2007). *Arabian sands.* London: Penguin Classics (originally published 1959).

Thornton, S.N. (2010). Thirst and hydration: Physiology and consequences of dysfunction. *Physiology and Behavior, 100*, 15–21.

Tilburg, W.A.P. and Igou, E.R. (2012). On boredom: Lack of challenge and meaning as distinct boredom experiences. *Motivation and Emotion, 36*, 181–94.

Todman, M. (2003). Boredom and psychotic disorders: Cognitive and motivational issues. *Psychiatry, 66*, 146–67.

Tomalin, N. and Hall, R. (2003). *The strange last voyage of Donald Crowhurst.* London: Hodder and Stoughton.

Totterdell, P. et al. (1998). Evidence of mood linkage in work groups. *Journal of Personality and Social Psychology, 74*, 1504–15.

Travis, F. et al. (2002). Patterns of EEG coherence, power, and contingent negative variation characterize the integration of transcendental and waking states. *Biological Psychology, 61*, 293–319.

Trunnell, E.P. et al. (1996). Optimizing an outdoor experience for experiential learning by decreasing boredom through mindfulness training. *Journal of Experiential Education, 19*, 43–9.

Tsai, L.L. et al. (2005). Impairment of error monitoring following sleep deprivation. *Sleep, 28*, 707–13.

Tumbat, G. and Belk, R.W. (2011). Marketplace tensions in extraordinary experiences. *Journal of Consumer Research, 38*, 42–61.

Twenge J.M. et al. (2001). If you can't join them, beat them: Effects of social exclusion on aggressive behavior. *Journal of Personality and Social Psychology, 81*, 1058–69.

Ulrich, R.S. et al. (1991). Stress recovery during exposure to natural and urban environments. *Journal of Environmental Psychology, 11*, 201–30.

Urbina, D.A. and Charles, R. (2014). Enduring the isolation of interplanetary travel: A personal account of the Mars500 mission. *Acta Astronautica, 93*, 374–83.

Vachon-Pressau, E. et al. (2013). Acute stress contributes to individual differences in pain and pain-related brain activity in healthy and chronic pain patients. *The Journal of Neuroscience, 33*, 6826–33.

Vaernes, R.J. (1993). *EXEMSI '92 executive summary. (NUTEC Report 16-03)*. Paris: European Space Agency.

Vallerand, R.J. (2008). On the psychology of passion: In search of what makes people's lives most worth living. *Canadian Psychology/Psychologie Canadienne, 49*, 1–13.

Vallerand, R.J. (2012). From motivation to passion: In search of the motivational processes involved in a meaningful life. *Canadian Psychology/Psychologie Canadienne, 53*, 42–52.

Vallerand, R.J. et al. (2003). Les passions de l'ame: On obsessive and harmonious passion. *Journal of Personality and Social Psychology, 85*, 756–67.

Van Dierendonck, D. and Te Nijenhuis, J. (2005). Flotation restricted environmental stimulation therapy (REST) as a stress-management tool: A meta-analysis. *Psychology and Health, 20*, 405–12.

Van Schaik, V. (2008). *Fatally flawed: The quest to be deepest*. South Africa: Liquid Edge Publishing.

Vgontzas, A.N. et al. (2007). Daytime napping after a night of sleep loss decreases sleepiness, improves performance, and causes beneficial changes in cortisol and interleukin-6 secretion. *American Journal of Physiology—Endocrinology and Metabolism, 292*, E253–61.

Vickers, K. et al. (2012). The 35% carbon dioxide test in stress and panic research: Overview of effects and integration of findings. *Clinical Psychology Review, 32*, 153–64.

Viesturs, E. and Roberts, D. (2006). *No shortcuts to the top*. New York: Broadway Books.

Viesturs, E. and Roberts, D. (2009). *K2: Life and death on the world's most dangerous mountain*. New York: Broadway Books.

Viscusi, W.K. (1984). The lulling effect: The impact of child-resistant packaging on aspirin and analgesic ingestions. *The American Economic Review, 74*, 324–7.

Vodanovich, S.J. (2003). Psychometric measures of boredom: A review of the literature. *The Journal of Psychology, 137*, 569–95.

Wagner, A.M. and Houlihan, D.D. (1994). Sensation seeking and trait anxiety in hang-glider pilots and golfers. *Personality and Individual Differences, 16*, 975–7.

Wagner, M. (2001). Behavioral characteristics related to substance abuse and risk-taking, sensation seeking, anxiety, sensitivity, and self-reinforcement. *Addictive Behaviors, 26*, 115–20.

Walford R.L., Mock, D., and Verdery, R. (2002). Caloric restriction in Biosphere 2: Alterations in physiologic, hematologic, hormonal, and biochemical parameters in humans restricted for a 2-year period. *Journal of Gerontology, 57A*, B211–24.

Wall, P. (2000). *Pain: The science of suffering*. London: Phoenix.

Wan, R. et al. (2010). Cardioprotective effect of intermittent fasting is associated with an elevation of adiponectin levels in rats. *Journal of Nutritional Biochemistry, 21*, 413–17.

Wanberg, C.R. (2012). The individual experience of unemployment. *Annual Review of Psychology, 63*, 369–96.

Ward, N. and O'Brien, S. (2007). *Left for dead*. London: A & C Black.

Watt, J.D. and Vodanovich, S.J. (1992). Relationship between boredom proneness and impulsivity. *Psychological Reports, 70*, 688–90.

Watt, J.D. and Vodanovich, S.J. (1999). Boredom proneness and psychosocial development. *The Journal of Psychology*, *133*, 303–14.

Weinstein, N.D. (1978). Individual differences in reactions to noise: A longitudinal study in a college dormitory. *Journal of Applied Psychology*, *63*, 458–66.

Westhoff, J.L., Koepsell, T.D., and Littell, C.T. (2012). Effects of experience and commercialisation on survival in Himalayan mountaineering: Retrospective cohort study. *British Medical Journal*, *344*, e3782–e3782.

Wetherell, M.A. et al. (2006). The four-dimensional stress test: Psychological, sympathetic-adrenal-medullary, parasympathetic and hypothalamic-pituitary-adrenal responses following inhalation of 35% CO_2. *Psychoneuroendocrinology*, *31*, 736–47.

Wetzler, B. (2001). Base Camp confidential: An oral history of Everest's endearingly dysfunctional village. *Outside Magazine, April 2001*.

Weybrew, B.B. (1991). Three decades of nuclear submarine research: Implications for space and Antarctic research. In A.A. Harrison, Y.A. Clearwater, and C.P. McKay (Eds) *From Antarctica to outer space: Life in isolation and confinement* (pp. 103–14). New York: Springer-Verlag.

Wheeler, S. (1997). *Terra incognita: Travels in Antarctica*. London: Vintage.

Whiteclay Chambers III, J. (2010). Office of Strategic Services training during World War II. *Studies in Intelligence*, *54*, 1–27.

Wickwire, J. and Bullitt, D. (1998). *Addicted to danger*. New York: Pocket Books.

Wilde, G.J., Robertson, L.S., and Pless, I.B. (2002). Does risk homoeostasis theory have implications for road safety? *British Medical Journal*, *324*, 1149–52.

Wilkinson, A. (2012). *The ice balloon*. London: Fourth Estate.

Williams, K.D. (2007). Ostracism. *Annual Review of Psychology*, *58*, 425–52.

Williams, K.D., Cheung, C.K., and Choi, W. (2000). Cyberostracism: Effects of being ignored over the Internet. *Journal of Personality and Social Psychology*, *79*, 748–62.

Williams, K.D. and Nida, S.A. (2011). Ostracism: Consequences and coping. *Current Directions in Psychological Science*, *20*, 71–5.

Williams, K.D. and Sommer, K.L. (1997). Social ostracism by coworkers: Does rejection lead to loafing or compensation? *Personality and Social Psychology Bulletin*, *23*, 693–706.

Williamson, A. et al. (2011). The link between fatigue and safety. *Accident Analysis* and *Prevention*, *43*, 498–515.

Williamson, A.M. and Feyer, A-M. (2000). Moderate sleep deprivation produces impairments in cognitive and motor performance equivalent to legally prescribed levels of alcohol intoxication. *Occupational and Environmental Medicine*, *57*, 649–55.

Wise, J. (2009). *Extreme fear*. London: Palgrave Macmillan.

Wolfe, T. (2005). *The right stuff*. London: Vintage.

Wu, G.J. (2013). The dark side of adventure: Exploring the stress-coping strategies of mountaineers' significant others regarding high altitude mountaineering expeditions. *Applied Research in Quality of Life*, *8*, 449–65.

Yamagishi, T. (2001). Trust as a form of social intelligence. In K.S. Cook (Ed.) *Trust in Society* (pp. 121–47). New York: Russell Sage Foundation.

Yan, X.W. and England, M.E. (2001). Design evaluation of an Arctic research station from a user perspective. *Environment and Behavior*, *33*, 449–70.

Yates, J.F. (2001). 'Outsider': Impressions of naturalistic decision making. In E. Salas and G. Klein (Eds) *Linking expertise and naturalistic decision making* (pp. 9–33). Mahwah, NJ: Lawrence Erlbaum Associates.

Yeomans, M.R. (1998). Taste, palatability and the control of appetite. *Proceedings of the Nutrition Society, 57,* 609–15.

Yerkes, R.M. and Dodson, J.D. (1908). The relation of strength of stimulus to rapidity of habit formation. *Journal of Comparative and Neurological Psychology, 18,* 459–82.

Yoshida, W. et al. (2013). Uncertainty increases pain: Evidence for a novel mechanism of pain modulation involving the periaqueductal gray. *The Journal of Neuroscience, 33,* 5638–46.

Zeidan, F. et al. (2010a). Mindfulness meditation improves cognition: Evidence of brief mental training. *Consciousness and Cognition, 19,* 597–605.

Zeidan, F. et al. (2010b). The effects of brief mindfulness meditation training on experimentally induced pain. *Journal of Pain, 11,* 199–209.

Zeidan, F. et al. (2011). Brain mechanisms supporting the modulation of pain by mindfulness meditation. *The Journal of Neuroscience, 31,* 5540–8.

Zorpette, G. (1997). A real dive. *Scientific American, 277,* 32–6.

Zorpette, G. (1999). Extreme sports, sensation seeking and the brain. *Scientific American, 10,* 56–9.

Zubek, J.P. (1973). Behavioral and physiological effects of prolonged sensory and perceptual deprivation: A review. In J.E. Rasmussen (Ed.) *Man in isolation and confinement* (pp. 9–84). Chicago: Aldine Publishing Company.

Zuckerman, M. (1969). Variables affecting deprivation results. In J.P. Zubek (Ed.) *Sensory deprivation: Fifteen years of research.* New York: Appleton-Century-Crofts.

Zuckerman, M. (1994). *Behavioral expressions and biosocial bases of sensation seeking.* Cambridge: Cambridge University Press.

Zuckerman, M. (2006). *Sensation seeking and risky behavior.* Washington, DC: American Psychological Association Press.

INDEX

PERSONALITY

What makes you the way you are

Daniel Nettle

978-0-19-921143-2 | Paperback | £8.99

Why are some people worriers, and others wanderers? Why do some people seem good at empathizing, and others at controlling? We have something deep and consistent within us that determines the choices we make and the situations we bring about. Daniel Nettle takes the reader on a tour through the science of human personality, introducing the five 'dimensions' on which every personality is based, and using an unusual combination of individual life stories and scientific research. Showing how our personalities stem from our biological makeup, Nettle looks at the latest findings from genetics and brain science, considers the evolutionary origins and consequences of personality variation, and even includes a questionnaire for you to assess your own personality against the five dimensions.

Sign up to our quarterly e-newsletter **http://academic-preferences.oup.com/**

BAD MOVES

How decision making goes wrong,
and the ethics of smart drugs

Barbara Sahakian and Jamie Nicole LaBuzetta

978-0-19-966847-2 | Hardback | £14.99

"With this accessible primer, full of medical anecdotes and clear explanations, Sahakian and Labuzetta prepare the public for an informed discussion about the role of drugs in our society."

Nature

The realization that smart drugs can improve cognitive abilities in healthy people has led to growing general use, with drugs easily available via the Internet. Sahakian and Labuzetta raise ethical questions about the availability of these drugs for cognitive enhancement, in the hope of informing public debate about an increasingly important issue.

SOCIAL

Why our brains are wired to connect

Matthew D. Lieberman

978-0-19-964504-6 | Hardback | £18.99

"This isn't just fascinating for its own sake. Lieberman has a social and political purpose."

Julian Baggini, *Financial Times*

"*Social* is the book I've been waiting for: a brilliant and beautiful exploration of how and why we are wired together, by one of the field's most prescient pioneers."

Daniel Gilbert, Harvard University

Why are we influenced by the behaviour of complete strangers? Why does the brain register similar pleasure when I perceive something as 'fair' or when I eat chocolate? Why can we be so profoundly hurt by bereavement? The young discipline of 'social cognitive neuroscience' has been exploring this fascinating interface between brain science and human behaviour since the late 1990s.

Sign up to our quarterly e-newsletter **http://academic-preferences.oup.com/**

BRAINWASHING

The science of thought control

Kathleen Taylor

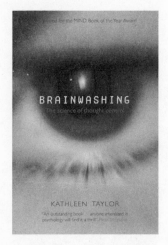

978-0-19-920478-6 | Paperback | £9.99

"An ambitious and well-written study"

The Guardian

In *Brainwashing*, Kathleen Taylor brings the worlds of neuroscience and social psychology together for the first time. In elegant and accessible prose, and with abundant use of anecdotes and case-studies, she examines the ethical problems involved in carrying out the required experiments on humans, the limitations of animal models, and the frightening implications of such research. She also explores the history of thought-control and shows how it still exists all around us, from marketing and television, to politics and education.

FUTURE SCIENCE

Essays from the cutting edge

Edited by Max Brockman

978-0-19-969935-3 | Paperback | £9.99

"Punchy, provocative and packed with fascinating insights" *BBC Focus magazine*

"Marvellous" *Independent on Sunday*

The next wave of science writing is here. Editor Max Brockman has talent-spotted 19 young scientists, working on leading-edge research across a wide range of fields. Nearly half of them are women, and all of them are great communicators: their passion and excitement makes this collection a wonderfully invigorating read. *Future Science* covers a variety of issues from neuroscience and evolutionary psychology to plant populations and the new age of oceanography.

Sign up to our quarterly e-newsletter **http://academic-preferences.oup.com/**

COLLIDING CONTINENTS

*A geological exploration of the Himalaya,
Karakoram, and Tibet*

Mike Searle

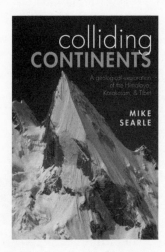

978-0-19-965300-3 | Hardback | £25.00

"There's something here to please anyone on the geology spectrum: the student wanting to understand how the fundamentals are applied; the academic intrigued by the science; the climber dreaming of virgin territory. All can learn from the master in this excellent book."

Simon Cook, *Oman Daily Observer*

The crash of the Indian plate into Asia is the biggest known collision in geological history, and it continues today. The result is the Himalaya and Karakoram - one of the largest mountain ranges on Earth. In this beautifully illustrated book, Mike Searle, a geologist at the University of Oxford and one of the most experienced field geologists of our time, uses his personal accounts of extreme mountaineering and research in the region to piece together the geological processes that formed such impressive peaks.

Sign up to our quarterly e-newsletter **http://academic-preferences.oup.com/**